Topics in Current Physics

23

Topics in Current Physics Founded by Helmut K. V. Lotsch

1 **Beam-Foil Spectroscopy**
Editor: S. Bashkin

2 **Modern Three-Hadron Physics**
Editor: A. W. Thomas

3 **Dynamics of Solids and Liquids by Neutron Scattering**
Editors: S. W. Lovesey and T. Springer

4 **Electron Spectroscopy for Surface Analysis**
Editor: H. Ibach

5 **Structure and Collisions of Ions and Atoms**
Editor: I. A. Sellin

6 **Neutron Diffraction**
Editor: H. Dachs

7 **Monte Carlo Methods** in Statistical Physics
Editor: K. Binder

8 **Ocean Acoustics**
Editor: J. A. DeSanto

9 **Inverse Source Problems** in Optics
Editor: H. P. Baltes

10 **Synchrotron Radiation**
Techniques and Applications
Editor: C. Kunz

11 **Raman Spectroscopy** of Gases and Liquids
Editor: A. Weber

12 **Positrons in Solids**
Editor: P. Hautojärvi

13 **Computer Processing of Electron Microscope Images**
Editor: P. W. Hawkes

14 **Excitons** Editor: K. Cho

15 Physics of **Superionic Conductors**
Editor: M. B. Salamon

16 **Aerosol Microphysics I**
Particle Interactions
Editor: W. H. Marlow

17 **Solitons**
Editors: R. Bullough, P. Caudrey

18 **Magnetic Electron Lens Properties**
Editor: P. W. Hawkes

19 Theory of **Chemisorption**
Editor: J. R. Smith

20 **Inverse Scattering Problems** in Optics
Editor: H. P. Baltes

21 **Coherent Nonlinear Optics**
Recent Advances
Editors: M. S. Feld and V. S. Letokhov

22 **Electromagnetic Theory of Gratings**
Editor: R. Petit

23 **Structural Phase Transitions I**
Editors: K. A. Müller and H. Thomas

24 **Amorphous Solids**
Low-Temperature Properties
Editor: W. A. Phillips

25 **Mössbauer Spectroscopy II**
The Exotic Side of the Effect
Editor: U. Gonser

26 **Crystal Cohesion and Conformational Energies**
Editor: R. M. Metzger

Structural Phase Transitions I

Edited by K. A. Müller and H. Thomas

With Contributions by
B. Dorner P. A. Fleury B. Lüthi K. Lyons
K. A. Müller W. Rehwald

With 61 Figures

Springer-Verlag Berlin Heidelberg New York 1981

Professor Dr. K. Alex Müller
IBM, Zürich Research Laboratory, CH-8803 Rüschlikon, Switzerland

Professor Dr. Harry Thomas
Institut für Physik, Universität Basel, CH-4056 Basel, Switzerland

ISBN-13: 978-3-642-81533-1 e-ISBN-13: 978-3-642-81531-7
DOI: 10.1007/978-3-642-81531-7

Library of Congress Cataloging in Publication Data. Main entry under title: Structural phase transitions. (Topics in current physics ; v. 23). Bibliography: p. Includes index. 1. Phase transformations (Statistical physics) 2. Solid state physics. 3. Order-disorder models. I. Müller, K. Alex, 1927– II. Thomas, Harry, III. Dorwer, Bruno. IV. Series.
QC176.8.P45S77 530.4'1 80-23544

Softcover reprint of the hardcover 1st edition 1981

2153/3130-543210

Preface

The growth and maturity of research in structural phase transitions (SPT) make it an appropriate subject for the Topics in Current Physics series. The maturing process is, however, by no means complete. New areas such as incommensurable SPT, quasi-low-dimensional systems, systems containing lattice disorder due to impurities or as mixed crystals, multicritical points, and quantum effects have recently come under focus. The understanding of the dynamics, be it microscopic soft-mode theory or critical dynamics, more specifically the central-peak problem, is also still quite incomplete. On the other hand, there are areas which are genuinely consolidated. On the theoretical side, these concern symmetry properties, Landau theory, and the application of static renormalization theory to critical phenomena. Also, the use of various complementary experimental techniques, with their specific merits, are well in hand.

The field of STP's and of the various methods of investigation range so widely that it appeared appropriate to invite a number of scientists to review their respective areas of expertise to which they have made significant contributions. Therefore, the style and taste in the different chapters will, of course, vary to some extent. This diversity, however, guarantees a penetration of each area, in width and depth, to a degree which would have been difficult to achieve with a single author or through a small team covering the whole field. For instance, there are very few experimentalists who can cover scattering techniques as well as magnetic resonance methods with a comparable degree of competence, although important contributions to the field have been obtained by both techniques.

This volume is the first of an intended total of three. In the Introduction following this Preface the more general aspects of our content will be emphasized. This volume contains three chapters on experimental research. The contribution by Lyons and Fleury (Chap.2) discusses optical investigations, particularly light scattering, followed by an extensive bibliography. Then, Dorner reviews the results achieved by inelastic neutron scattering (Chap.3). He, together with Comes, has already presented certain aspects of this area, especially of a technical nature, in an earlier article in Topics in Current Physics (Vol.3). The new chapter emphasizes results not touched upon in the previous one. Chapter 4 by Lüthi and Rehwald gives an up-to-date account of the ultrasonic work. Another experimental volume is planned to consist again of three chapters, including results achieved with the remaining experimental techniques used in SPT, electron-paramagnetic and nuclear magnetic resonance, dielectric measurements, and caloric techniques. The theoretical volume will start with dynamic lattice theory, then go on to the Landau theory, general symmetry properties, and renormalization group theory, and end with a special chapter on the Jahn-Teller induced SPT.

It is the hope of the undersigned that the efforts made may serve their purpose, to help as an introduction into the various area of research in SPT, give a valid description of the state of the art as well as its possible future lines of research, and to serve as a reference.

Zürich, Basel, October, 1980 *K.A. Müller. H. Thomas*

Contents

List of Contributors

Dorner, Bruno

Institut Laue-Langevin B.P. 156, Centre de Tri, Avenue des Martyrs, F-38042 Grenoble, France

Fleury, Paul A.

Bell Laboratories, Murray Hill, N.J. 07974, USA

Lüthi, Bruno

Physikalisches Institut, Johann Wolfgang Goethe Universität, D-6000 Frankfurt/Main, Fed. Rep. of Germany

Lyons, Ken

Bell Laboratories, Murray Hill, N.J. 07974, USA

Müller, K. Alex

IBM, Zürich Research Laboratory, CH-8803 Rüschlikon, Switzerland

Rehwald, Walther

Laboratories RCA Ltd., CH-8048 Zürich, Switzerland

1. Introduction

K. A. Müller

With 2 Figures

The areas of research on structural phase transitions (SPT) enumerated in the Preface give a general impression of the field of interest addressed in this series. Here we want to provide a wider perspective. The change of structure at a phase transition in a solid can occur in two quite distinct ways. First of all, those transitions where the atoms of a solid *reconstruct* a new lattice, for example, when graphite transforms into diamond or if an amorphous solid changes to the crystalline state. Secondly, there are those where a regular lattice is *distorted* slightly "without in any other way disrupting the linkage of the net" according to BUERGER [1.1]. This can occur as a result of small *displacements* in the lattice position of single atoms or molecular units on the one hand or the *ordering* of atoms or molecules among various equivalent positions on the other hand. Due to the matter transport, which is inherently involved in reconstructive transitions, they are often slow (recrystallization). Because they are transitions which are not symmetry related in any way they have to be of first order. Reconstructive transitions are not considered in the following chapters. It is only the second category, above, to which this series is devoted. Therefore SPT are understood throughout in this narrower sense, introduced a decade ago [1.3] and used widely since then by solid-state physicists but not among crystallographers [1.2]. Martensitic phase transformations are also somewhat outside the present scope and are addressed only in Chap.4.

In the study of phase transitions the order parameter $\eta(T)$ is a crucial quantity. Below the transformation T_c it is nonzero and increases on cooling. In a ferromagnet it is the mangetization $M(T) = V^{-1} \sum_i (\mu^i_{0+} - \mu^i_{0-})$ which measures the number of ordered atomic magnetic moments μ_0, per unit volume, of an infinite solid, in the magnetization direction. This corresponds to the order parameter $Q(T) = V^{-1} \sum_i (Q^i_{0+} - Q^i_{0-})$ in a pure order-disorder SPT. A distinction between the latter and a displacive transition is possible on the basis of atomic single cell potentials [1.4]. This is shown in Fig.1.1, schematically for one spatial coordinate Q: with anharmonic potential $V(Q) = aQ^2 + bQ^4$ with constants $a < 0$ and $b > 0$. This double well potential has two minima and a maximum, their difference ΔE being $\Delta E = a^2/4b$. If the depth of the two energy wells $\Delta E \gg kT_c$ the transition occurs due to ordering between $Q^i_{0\pm} = \pm\sqrt{-a/2b}$. On the other hand, if $\Delta E \ll kT_c$ a *continuous* cooperative displacement of atoms along Q below T_c is found as a function of temperature, at least in mean field theory [1.5].

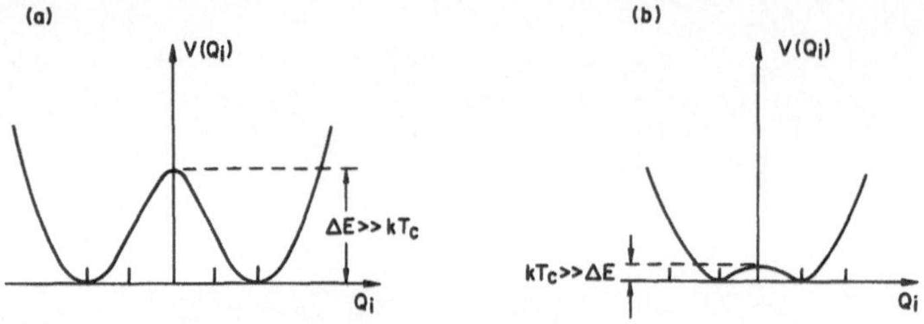

Fig. 1.1. Single cell potentials in (a) order-disorder (b) displacive SPT system

The transitions themselves result from the coupling of different cells, i and j, in the regular lattice; there is an interaction energy $V = \sum_{i>j} c_{ij}Q_iQ_j$. A nonzero single cell anharmonicity is, however, also needed in displacive systems. A completely harmonic potential, i.e., a parabolic single cell potential, does not yield a phase transition despite $V \neq 0$.

Using self-consistent mean field theory one derives the free energy of the system to be [1.5]

$$F = F_0 + A\langle Q\rangle^2 + B\langle Q\rangle^4 + \text{higher order} , \quad (1.1)$$

where F_0 contains all other degrees of freedom of the system, $\langle Q\rangle = Q(T)$, $A = \alpha(T - T_c)$ and α and B can be viewed as nearly temperature independent, near T_c. Equation (1.1) is the expression Landau found from group theoretical arguments in 1937, assuming that the space group of the low-temperature phase is a subgroup of its high symmetry parent or prototype phase [1.6], and no third-order invariant exists. Furthermore, the order parameter Q(T) is a scalar; it has but one dimension. We shall come back to the fundamental importance of the Landau free energy expansion for the more general cases in SPT, as well as its modern Hamiltonian version in renormalization group theory.

Upon minimization of (1.1) with respect to $\langle Q\rangle$, one finds a second-order transition at T_c when $B > 0$ with a continuous order parameter varying as $\langle Q\rangle^2 \propto (T_c - T)$ (see Fig.1.2). If symmetry allows a third-order invariant $C\langle Q\rangle^3$ in (1.1), a first-order transition at $T_0 > T_c$ results with a discontinuous jump of the order parameter [1.7]. The transition is then sometimes called "weak" first order not only because the jump in $\langle Q\rangle$ can be quite small, but because the low-temperature phase is still a subgroup of the high-temperature one. On the other hand, for the reconstructive transitions mentioned at the beginning, no such symmetry relation exists. The transition occurs, in that case, when the high and low symmetry free enthalpies equal each other, and there can be considerable metastability.

The dynamic behavior of the two types of distortive SPT is also quite different. The order-disorder systems behave like the magnetic ones. Above T_c their excitation

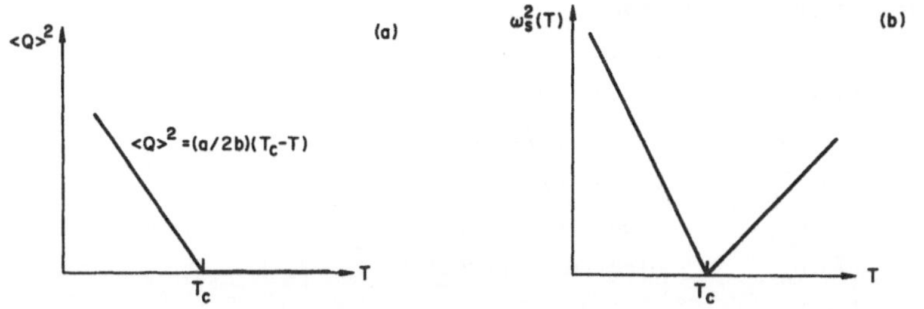

Fig. 1.2. (b) $\omega_s^2 = 2(a/m)(T-T_c)$ for $T>T_c$; $\omega_s^2 = 4(a/m)(T_c-T)$ for $T<T_c$

spectrum shows relaxational character and is centered around $\omega = 0$. Only below T_c do we find a mode at finite frequency $\omega \neq 0$ (as for spin waves in ferromagnets). On the other hand, in displacive systems, a mode of finite frequency exists even above T_c, and tends to freeze out on approaching T_c from above. It is obtained from (1.1) by solving the dynamical matrix [1.5]

$$\partial^2 F/\partial\langle Q\rangle^2 = m\omega_s^2(T) \ , \qquad (1.2)$$

taking into account the inertia of the atoms.

This soft mode $\omega_s(T)$ was a concept introduced two decades ago first by COCHRAN [1.8,9] and was extremely fruitful both for experimental and theoretical research in SPT. Table 1.1 lists the SPT systems and their soft mode character. As ferroelectric examples we cite $BaTiO_3$ which is more of the displacive, and KH_2PO_4 which is more of the order-disorder variety, although it is now recognized that they are by no means "pure" examples. Using the terms of Fig. 1.1, ΔE is of the order of kT_c in both cases, in one somewhat larger and in the other slightly smaller. The soft-mode concept has been extended to pure order-disorder relaxational systems [1.10], where in a mean field approximation $2\pi/\tau = \omega_r \propto (T-T_c)$, whereas $\omega_s \propto (T-T_c)^{1/2}$ (see Fig.1.2). However, if the soft phonon is overdamped its temperature dependence cannot be distinguished from a relaxational response [1.11]. The Jahn-Teller systems listed in Table 1.1 also deserve comment. These phase transitions come about through the coupling of electronic and acoustic degrees of freedom. The former can be localized like the 4f electrons in $TbVO_4$, or else delocalized as in Nb_3Sn, in which case they are called band JT transitions.

The modes of a lattice, or their quantized equivalent, the phonons, are probed most directly by inelastic scattering of photons or neutrons. Chapter 2 in the present volume is devoted mainly to the first of these techniques, light scattering, and Chap.3 is devoted entirely to neutron scattering. Fleury and Lyons give characteristic examples of the various SPT systems in each section of their chapter. Their choices, for particular substances, differ in part from the list in Table 1.1. (If the light is scattered by optical modes it is called Raman scattering whereas

Table 1.1. SPT systems and their soft modes

short range forces	($SrTiO_3$)	optic (xy)
dipolar, ferroelectric	($BaTiO_3$)	t. optic (x)
	(KH_2PO_4)	t. optic/t. acoustic
ferroelastic	(TcO_2)	acoustic
Jahn-Teller, ion	($TbVO_4$)	electronic/acoustic
band	(Nb_3Sn)	
charge density wave	($NbSe_2$)	electronic/acoustic

that scattered from acoustic modes is termed Brillouin scattering [1.7].) Their chapter gives the reader an understanding of the different SPT types and also of characteristic differences between possible soft modes. Furthermore, examples of Brillouin center, boundary, and incommensurate soft modes are reviewed (we shall define these terms below). Their latter section is a valuable introduction to this type of SPT.

The modes of a crystal, characterized by a dispersion relationship $\omega(\underline{q})$, of a crystal, are collective excitations and extend over the whole wave vector space $\underline{q}$ of the Brillouin zone. At a displacive transition the freezing out of the soft mode produces the lattice found on the low-temperature side of the transition. For $T > T_c$, $\omega_s(\underline{q})$ has a minimum at some wave vector in $\underline{q}$ space. This minimum is most often found at the zone center $\underline{q} = 0$ or at particular points of the Brillouin zone boundary $\underline{q}_b$. Accordingly they are termed *zone center* and *zone boundary transitions* in Chap. 2. In the latter case the unit cell gets doubled (or subject to even higher multiplication) in the low-temperature phase because the point $\underline{q}_b$ gets "folded into" the zone center. (The soft mode unit vector in the low-temperature phase transforms as the identity representation of the space group). Denoting the unit cell volumes by z^h and z^ℓ, respectively, one can describe the situation statically by

$$z^\ell = nz^h \tag{1.3}$$

with n = 1 and n = 2r (r integer) in the two cases. These cases have been termed *ferro* and *antiferrodistortive* by GRÄNICHER and MÜLLER [1.12] earlier. In certain crystals molecular units order, as does NH_4 in NH_4Cl, below the cubic-cubic ordering phase transformation. Because no distortion occurs at T_c, the transition is called *ferro-order* (see Chap.3 for a discussion). The corresponding antiferro order in NH_4Br is equivalent to antiferrodistortive as the unit cell is doubled and distorted. As long as n in (1.3) is a rational number $n = r_\alpha/r_\beta$ with integral r_i and $r_\alpha > r_\beta$, the corresponding wave vector $\underline{q}$ of $\omega_s(\underline{q})$, inside the zone, is a rational multiple of a particular Brillouin zone boundary wave vector $\underline{q}_b$. The transition is then called *commensurate*. If however n is an irrational number the transitions are called *incommensurate* [1.13]. Some of the most recent studies in SPT are aimed towards an understanding of this interesting class of incommensurate transitions.

Up to now one may think that the descriptions of SPTs in terms of free energy or of soft modes are equivalent. We want to point out here that the free energy discussion is more general for the following reasons: First, if a transition has essentially reorientational and/or relaxational character, as does $NaNO_2$ where electric NO_2 dipoles reorient and order ferroelectrically, many solid-state physicists are reluctant to name this a zone center mode transition. The above-mentioned NH_4Cl is another example. A more important and second reason: There are continuous transitions where no microscopic dynamical fluctuations, i.e., soft or relaxational modes of finite wave vector, are present. A crystal has, in general, 6 independent elastic degrees of freedom, but only 3 acoustic modes. If one of these acoustic modes is soft, the transition is a ferroelastic SPT (see Table 1.1). However, one of the other 3 independent elastic degrees, not related to the acoustic modes, may become critical: then a ferroelastic transition occurs but no acoustic mode becomes soft. It is just an isolated point at the zone center whose elastic constant shows temperature dependence. Then the crystal free energy is given exactly by (1.1). Such a "type zero" SPT occurs generally for symmetry nonbreaking transitions and has most recently been observed in KH_2PO_4, in the presence of an electric field [1.14].

The wavelength of light is of a few thousand Å; this limits the wave vector scattering range of phonons essentially to the center of the Brillouin zone. On the other hand, the wavelength of thermal neutrons is of the order of Angstroms and their energy matches that of phonons [1.7]. This allows probing over the entire Brillouin zone. In Chap.3, Dorner first recalls the general aspects of inelastic neutron scattering near SPT and soft mode analysis. He then goes on to review some of the classic work on displacive SPT. His next section is devoted to the observation of central peaks, i.e., critically enhanced excitations near $\omega = 0$, observed for temperatures T above T_c, first reported in $SrTiO_3$ [1.15], and distinct from the underdamped soft phonons, and thus not contained in the classical picture. The narrowness of the observed features clearly points to the involvement of impurities in the dynamics, but not necessarily to the critical enhancement of the central peak, as well as the nonfreezing out of the soft mode [1.16]. Other central peaks seen for $T \le T_c$ have been probed by light scattering and are reviewed in Chap.2. Both accounts bring in touch with the recent as yet in part unsolved problems as to the origin of these excitations.

Subsequently Dorner introduces and describes studies in molecular crystals and the occurring SPT. This is the first summary, in the literature, of the field. The importance of the incoherent structure factor for the determination of molecular reorientation rates is emphasized. Then, examples of the librational dynamics of molecules, and of their various degrees of coupling to the translational modes, if any, are given. The ensuing molecular cluster formation near T_c is discussed. The concise nature of the account also makes it useful for experts in SPT. Dorner's last section is devoted to incommensurable SPT. It is the first account of inelastic scattering results and can be recommended as an introduction as well as an update to what has been achieved in charge density wave systems as well as in insulators or more exotic quasi-one-dimensional chain systems.

The soft modes can couple to other modes in the lattice. If the soft mode is an optical one it can couple to acoustic modes. Thus Brillouin scattering from the latter can monitor the soft mode temperature dependence, damping, etc. Theoretical aspects of coupling are discussed in Chap.2. However, one can probe the acoustic mode directly with ultrasonic dispersion and absorption measurements. In Chap.4 Lüthi and Rehwald review in depth the work done with this technique. It is, in favorable cases, one of the most accurate in the entire field. The authors give a clear account of the theoretical analysis. Depending upon whether the strain in the crystal couples linearly, or through higher order, to the order parameter, the ultrasound behavior as a function of temperature can be quite different. In the first case the strain itself can be taken as the order parameter and an elastic constant vanishes at a second-order phase transition. Thoroughly explored examples of the ferroelectric-ferroelastic as well as the two Jahn-Teller varieties (see Table 1.1) are presented. This is followed by a review of those cases where the order parameter couples quadratically to strain, and a critical reduction of an elastic constant with enhanced ultrasound attenuation is observed, near T_c. They can be used to determine phase diagrams as a function of an external applied parameter. In short, their review covers all systems shown in Table 1.1, and also mixed valence systems and martensitic transitions, not discussed elsewhere in this series.

An SPT can be monitored by the local displacements of the constituents of the lattice. X-rays and elastic neutron scattering have long been standard techniques for such investigations. Results of this sort can be found in a number of works. We should like to mention here the well-balanced Chap.5 in MEGAW's book on crystal structures [1.2]. In the context of this series the understanding and use of space groups is obviously of importance; their relatively infrequent use may be traced to the fact that no good introductory text was available for solid-state physicists. This gap in the literature has now been filled with the indeed readable book by BURNS and GLASER [1.17]. Knowledge of group theory and space groups will, in part, be required to fully appreciate the two chapters in the theoretical volume on Landau theory and general symmetry properties.

There exist now a number of recent techniques for probing the existence, amount, and symmetry of local displacements in a solid. These are in part one to two orders of magnitude more sensitive than the older ones. Optical fluorescence is one technique which is reviewed in Chap.2. Electron and nuclear magnetic resonances (EPR and NMR) as well as the Mössbauer effect are others. The first is very sensitive to static displacements and will be presented in the second volume, as will NMR.

The latter is less sensitive but certainly probes the bulk of the crystal. It gives useful information about the local atomic or molecular time-dependent motion. Using these sensitive techniques, as well as ultrasound dispersion and absorption, it is now firmly established that the static and dynamic behavior near SPT which results from short range interactions, differs from the one predicted by Landau and soft mode theory in the absence of impurity effects. Furthermore, on application of external forces, such as uniaxial stress or hydrostatic pressure, multicritical points are observed in which two or more second-order phase boundaries join as the external fields are varied. An example is the bicritical point in $SrTiO_3$ mentioned in Chap.4. In this very active field renormalization group theory has predicted certain topologies which are universal and depend only on the symmetry of the lattice, its dimensionality, and that of the order parameter. Some were first verified in SPT, rather than in the "competing" field of magnetic transitions. Such findings and a chapter on dielectric and caloric methods are planned to be the content of the follow-on volume.

The various experimental techniques employed in SPT research are often complementary. Depending on the information wanted in a particular case and the samples available, one method may yield more accurate or valuable information than another. Often the availability of samples or their gross macroscopic properties may inhibit their investigation with a certain technique. For example, if the crystal is large, buth highly conducting, ultrasound measurements are easily performed but it is difficult to use paramagnetic resonance, because the microwaves cannot penetrate the sample. On the other hand, if the available crystals are very small but nonconducting EPR or light scattering can be well employed, whereas classical ultrasound absorption methods are excluded due to phase and amplitude inhomogeneities in the small sample. The new phonon echo method may become useful here in the future [1.18]. Small crystals also create difficulty for inelastic neutron scattering, even in high flux reactors, especially if atoms with small scattering cross section have to be probed.

The above examples show that an assessment of experimental methods is in order. This has been done in each of the chapters. The descriptions of the techniques are concise and to the point, but fairly complete in Chaps.2 and 4. In view of the fact that the classical inelastic neutron scattering techniques have already been described in [1.11], they have not been repeated in Chap.3. This gave space for a review of the most recent developments of high-resolution instruments.

Acknowledgement. The help of various people, especially R. Landauer, in improving the mansucript is greatly appreciated.

References

1.1 M.J. Buerger: In *Phase Transitions in Solids* (Wiley, New York 1951) p.133
1.2 H.D. Megaw: *Crystal Structures, A Working Approach* (W.B. Saunders, Philadelphia 1973)
1.3 K.A. Müller, W. Berlinger, F. Waldner: Phys. Rev. Lett. *21*, 814 (1968)
1.4 A.D. Bruce: In *Solitons and Condensed Matter Physics*, Springer Series in Solid-State Sciences, Vol. 8, ed. by A.R. Bishop, T. Schneider (Springer, Berlin, Heidelberg, New York 1978) p. 116
1.5 H. Thomas: In *Structural Phase Transitions and Soft Modes*, Proceedings of the Nato Advanced Study Institute, ed. by E.J. Samuelsen, E. Andersen, J. Feder (Universitetsforlaget, Oslo 1971)
1.6 L. Landau, E. Lifschitz: *Physique Statistique* (MIR, Moscow 1967)
1.7 C. Kittel: *Introduction to Solid State Physics*, 3rd ed. (Wiley, New York 1965)
1.8 W. Cochran: Phys. Rev. Lett. *3*, 412 (1959)
1.9 P.W. Anderson: In *Fizika dielektrikov*, ed. by G.I. Skanavi (Acad. Nauk, SSR, Moscow 1960) p. 290
1.10 R. Blinc, B. Zeks: In *Soft Modes in Ferroelectrics and Antiferroelectrics*, ed. by E. P. Wohlfarth (North-Holland, Amsterdam 1979)
1.11 B. Dorner, R. Comes: In *Dynamics of Solids and Liquids by Neutron Scattering*, ed. by S.W. Lovesey, T. Springer, Topics in Current Physics, Vol.3 (Springer, Berlin, Heidelberg, New York 1977) p.127
1.12 H. Gränicher, K.A. Müller: Mater. Res. Bull. *6*, 977 (1971)
1.13 A. Janner, T. Janssen: Phys. Rev. B*15*, 643 (1977)
1.14 E. Courtens, R. Gammon, S. Alexander: Phys. Rev. Lett. *43*, 1026 (1979)
1.15 T. Riste, E.J. Samuelsen, K. Otnes, J. Feder: Solid State Comm. *9*, 1455 (1971)
1.16 K.A. Müller: In *Dynamical Critical Phenomena and Related Topics*, Lecture Notes in Physics, Vol. 104, ed. by C.P. Enz (Springer, Berlin, Heidelberg, New York (1979) p. 210
1.17 G. Burns, A.M. Glazer: *Space Groups for Solid State Scientists* (Academic Press, New York 1978)
1.18 N.S. Shiren, R.L. Melcher: In *Phonon Scattering in Solids*, ed. by L.G. Hallis, W.U. Rampton, A.F.G. What (Plenum, New York 1976) p. 405

2. Optical Studies of Structural Phase Transitions

P. A. Fleury and K. Lyons

With 26 Figures

A review is presented of the application of optical techniques to the study of structural phase transitions. The techniques discussed include static measurements (refractive index, birefringence) as well as those providing dynamic information (light scattering, IR and optical absorption, fluorescence). The use of nonlinear optical techniques such as second harmonic generation is also discussed. In a background section, the various experimental techniques are delineated with descriptions of the relevant apparatus and theoretical considerations, including comparison to the information available from other techniques. The various classes of phase transitions are then discussed in turn, organized according to the symmetry and periodicity of the order parameter. Emphasis is on the developments of the last few years, including optical studies of central peak and mode coupling phenomena, and recent developments in the use of Fourier transform spectroscopy. An extensive bibliography on this very broad subject is included as an adjunct to the review, categorized by the type of phase transition and by the experimental technique involved.

2.1 Background

The recent rapid growth of the field of optical research, triggered by the advent of the laser, has been accompanied, not accidentally, by a comparable revolution in our understanding of phase transitions [2.1-3]. In this review we are concerned with the ways in which optical techniques have contributed to this understanding. There is a vast literature in this field, even within the last few years, and it would be impossible in the space of this review to describe it all in any reasonable detail. Therefore, we have decided to limit the discussion in several ways. We have chosen to exclude all work on plastic crystals. In addition, since most magnetic transitions do not involve structural changes and there have been two very recent review articles on optical investigation of magnetic crystals [2.4,5], we shall exclude these also from consideration. Finally, we have decided to present only certain specific examples in detail. This means that much of the literature is not referenced directly in the text. Thus, at the end of this review we include a categorized bibliography, representing the results of an extensive literature search. We feel that this organization best serves the purpose for which this review is intended.

In addition to the two mentioned above, other recent reviews relevant to our topic include a discussion of birefringence by COURTENS [2.6] and a review of the application of optical techniques to cooperative Jahn-Teller systems by GEHRING and GEHRING [2.7]. A review of the application of Brillouin scattering to structural phase transitions has been given by SMOLENSKY and HASHKOZEV [2.8]. The effects of pressure on light scattering spectra have been discussed by PEERCY [2.9]. Raman scattering near structural phase transitions has been reviewed by FLEURY [2.10], SCOTT [2.11], STEIGMEIER [2.12], and NAKAMURA [2.13]. Finally, the application of nonlinear optical techniques has been reviewed by VOGT [2.14]. In the present chapter we shall attempt to place more recent developments in perspective, and also to give a comprehensive picture of the contribution of optical techniques in general to the understanding of structural phase transitions.

There are a number of types of phase transitions to which optical techniques have been applied. It is beyond the scope of this review to delve into the details of the various microscopic mechanisms. Instead, we shall present, in each case, a simple introduction to the theory, with appropriate references to more detailed work for the interested reader. We shall concentrate on the experimental results and what they imply for the understanding of the phase transition. The remainder of this introductory section is devoted to a brief discussion of the properties of phase transitions in general and how they relate to the organization of the present chapter. The organization largely centers upon the symmetry types of the transitions. Thus, following a discussion of experimental techniques (Sect.2.2), we consider: ferroelastic transitions (Sect.2.3), nonpiezoelectric ferroelectrics (Sect.2.4), piezoelectric ferroelectrics (Sect.2.5), zone boundary transitions (Sect.2.6, improper ferroelectrics (Sect.2.7), and incommensurate transitions (Sect.2.8). These are followed by sections on nonlinear optics (Sect.2.9) and multicritical phenomena (Sect. 2.10). Section 2.11 records some concluding remarks, and a comprehensive bibliography is given at the end.

2.1.1 Phase Transitions in Solids: Definitions

The symmetry aspects of a solid-solid phase transition are fully described by the structures of the two phases [2.15]. The high-symmetry (usually high-temperature) phase is called the prototypic phase. At some temperature T_c the symmetry of the crystal changes. The low-symmetry phase is quantitatively described by the order parameter, ψ of the transition. If the transition occurs continuously (that is, $\psi \to 0$ smoothly) then the transition is said to be second order. Experimentally, one can only determine a phase transition to be "nearly second order" since it is always possible for a small jump in ψ or other properties to go undetected. This will be understood whenever we use the term "second order". Near a second-order phase transition, the crystal becomes "soft" with respect to the order parameter. More precisely, the corresponding susceptibility diverges and the fluctuations of ψ become very large and correlated over large spacetime volumes.

Let us now consider the translational symmetry of the order parameter. The form of the free energy in (2.1) below and the discussion of that equation serve as a precise definition of the order parameter. The spatial periodicity of the order parameter is not necessarily that of the prototypic phase. If it is, then the prototypic wave vector associated with the order parameter is $\underline{q}_\psi = 0$ and the phase transition is referred to as a *zone center* or *ferrodistortive* [2.16] transition. In a zone-center transition macroscopic quantities may couple linearly to the order parameter, such as spontaneous polarization (P_s for a ferroelectric transition) or spontaneous strain (X_s for a ferroelastic transition). If, on the other hand, the translational symmetry of the lattice changes at T_c, then ψ transforms as a nonzero wave vector of the prototypic phase ($\underline{q}_\psi \neq 0$) and ψ cannot couple linearly to a thermodynamic or macrosopic property. The best known examples of this behavior are the cell-doubling transitions of the perovskites such as $SrTiO_3$, $PrAlO_3$, etc. These are *zone boundary* transitions, sometimes referred to as *antiferrodistortive* [2.16]. A third and fundamentally different class of phase transitions, recognized only rather recently, is the class of *incommensurate* transitions, those in which $\underline{q}_\psi$ is not an integral submultiple of a translation vector of the prototypic phase [2.17]. Again in this case no linear coupling is allowed between ψ and macroscopic properties, but in addition *nonlinear* coupling of finite order is not allowed. Hence, in K_2SeO_4, which undergoes an incommensurate transition at T_i = 129 K, no spontaneous polarization (P_s) is observed, despite the polar character of the order parameter, until a "lock-in" transition occurs in which $\underline{q}_\psi$ discontinuously becomes a submultiple of a lattice vector thus allowing a (nonlinear) coupling to P_s.

The earliest theoretical attempt at explanation of the phase transition phenomenon has often been referred to as thermodynamic or mean field theory [MFT] [2.18]. In MFT, the correlated fluctuations are ignored and it is assumed that the crystal may be described solely in terms of its macroscopic (thermodynamic) properties. Fluctuations in these thermodynamic variables are then included by standard thermodynamic techniques. So long as the fluctuations on a scale larger than the interaction range in the crystal are not large, this is obviously a good approximation. However, in many systems the interactions are short range, and the fluctuation-dominated regime is readily accessible experimentally.

Despite this limitation MFT provides a useful background against which to view modern developments. Thus, we shall briefly sketch its development. The basic assumption is that the relevant portion of the free energy of the crystal may be expressed in terms of the average value of the order parameter $\langle\psi\rangle = \psi_0$,

$$A = a\psi_0^2 + b\psi_0^4 + c\psi_0^6 \ldots \quad . \tag{2.1}$$

That is, we assume

$$\langle\psi^n\rangle = \langle\psi\rangle^n \tag{2.2}$$

for all n. For a second-order transition, b is positive, and one usually assumes a linear dependence of a on temperature, $a = B(T - T_c)$. Analysis of the minima, $\psi = \psi_0$ in (2.1) then gives

$$\psi_0^2 \propto (T_c - T)^{2\beta} \; , \quad \chi_\psi \propto |T - T_c|^{-\gamma} \; ,$$

where χ_ψ is the susceptibility to a force conjugate to ψ. In MFT the critical exponents are given by $\gamma = 2\beta = 1.0$. The exponents β and γ have been introduced here in anticipation of the fact that they may experimentally deviate from the mean field values given above. This can be interpreted as a failure of the assumption (2.2). In fact, the extraction of "critical exponents" from soft mode spectra has often not been treated adequately. In addition to the obvious experimental difficulties associated with analysis of multicomponent spectra, particularly in the presence of mode coupling and background effects, there are some potential theoretical pitfalls of which one must be wary. These will be discussed further in Sect.2.2.3.

2.1.2 Symmetry Classification of Optical Techniques

Modern theories, in particular the recent renormalization group calculations, do arrive at a self-consistent picture of a phase transition and thus corrections to the MFT results above are found [2.19,20]. These theories are formulated in terms of order parameter correlation functions given by

$$\langle[\psi(0,0)]^n[\psi(\underline{r},t)]^m\rangle \; .$$

The most common of these correlation functions and the notations we shall use for them are

$$\begin{aligned} &\psi_0 \equiv \langle\psi\rangle \; ; \quad \langle\psi^2\rangle \equiv \langle\psi(\underline{r},t)\psi(\underline{r},t)\rangle \\ &G_1(\underline{r},t) \equiv \langle\psi(0,0)\psi(\underline{r},t)\rangle \\ &G_2(\underline{r},t) \equiv \langle[\psi(0,0)]^2[\psi(\underline{r},t)]^2\rangle \; , \end{aligned} \tag{2.3}$$

where the brackets signify averages over an appropriate ensemble. For convenience we shall refer to Fourier transforms by the following notation, for any quantity A, defined as a function of position r and t:

$$A(\underline{q},\omega) \equiv \int_{-\infty}^{\infty} d^3r dt A(\underline{r},t) \exp[i(\underline{q}\cdot\underline{r} + \omega t)] \; . \tag{2.4}$$

We now consider the way in which these correlation functions are related to optical experiments. Optical techniques fall into two categories on the basis of inversion symmetry. The matrix elements responsible for absorption and fluorescence (one-photon processes) transform as the vector components x,y, and z, while the polarization tensor components, which are relevant to light scattering and birefringence, transform as x^2, xy, etc. For materials which possess inversion symmetry it is clear that the order parameter cannot couple linearly to both sets of phenomena. From the standpoint of optical techniques, then, it is convenient to categorize phase transitions according to the existence of a center of symmetry. Higher order processes, described by tensors of higher rank, are not included in such a categorization, and will be discussed separately in the section on nonlinear optical techniques.

Table 2.1. Symmetry classification of optical techniques

Inversion symmetry? $T < T_c$	$T > T_c$	$q_\psi a$	Examples	Absorption fluorescence	Light scattering $S(q,\omega)$	Birefringence Δn
yes	yes	0	$PrAlO_3$	ψ_0, $\langle\psi^2\rangle$	G_1	ψ_0
yes	yes	$\pi/2$ inc,$\pi/3$	$SrTiO_3$ $TaSe_2$	ψ_0, $\langle\psi^2\rangle$	G_2	$\langle\psi^2\rangle$
no	yes	0 $\pi/2$ inc,$\pi/3$	$Pb_5Ge_3O_{11}$ $Gd_2(MbO_4)_3$ K_2SeO_4	ψ_0	G_2	$\langle\psi^2\rangle$
no	no	0 inc	KH_2PO_4 $BaMnF_4$	ψ_0	G_1	ψ_0

This categorization is displayed in Table 2.1, where we present the lowest order contribution of ψ to the position of spectral lines in fluorescence and optical absorption, to the scattered spectrum $S(\underline{q},\omega)$, and to the birefringence Δn, along with the examples we have chosen to represent each class in this review. We see that from the standpoint of optics there are four possible classes of phase transitions. In addition to inversion symmetry they are distinguished by the spatial periodicity of the order parameter. In the case where both phases are centrosymmetric, for a zone-center transition, strain must be an order parameter. Hence the fluctuations in ψ must be Raman active, which is to say that the linear term in the relation between the refractive index n and ψ is allowed. That is

$$n = n_0 + a_1\psi + a_2\psi^2 \quad . \tag{2.5}$$

If, on the other hand, the transition is not at $\underline{q} = 0$, then such a coupling is impossible regardless of inversion symmetry and the coefficient a_1 is zero. As we shall see in the next section, the scattered spectrum is given by the Fourier transform of the correlation function $\langle \delta n(\underline{r},t)\delta n(0,0)\rangle$ Hence, if $a_1 = 0$, the scattered spectrum is $G_2(\underline{q},\omega)$, but if $a_1 \neq 0$ it is given to lowest order by $G_1(\underline{q},\omega)$. In addition, if $a_1 \neq 0$, then there must be some component of the birefringence (Δn_{ij}) which is proportional to ψ_0 to lowest order. The quadratic term will also be present, of course, and may be observable. In the third group of lines of the table, ψ is of odd parity and, hence, the linear term in (2.5) must be zero, while in the fourth group of lines, it is allowed by symmetry. The corresponding lowest order behaviors of $S(\underline{q},\omega)$ and Δn are as shown.

On the other hand, the coupling of ψ to the *energy* of states involved in absorption and/or fluorescence (from or to the ground state, respectively) depends on the symmetry of both the *excited states* and the *order parameter*. If the *ion site* possesses a center of symmetry, these states must be of odd parity. Hence, if ψ is symmetric under inversion about the ion site, it cannot couple linearly to these states. But if ψ lacks inversion symmetry this coupling is in general allowed, as in the third group of lines of the table, the paraelectric-ferroelectric class of transitions. In a lattice which does not have inversion symmetry, no ion site can have it either, and, hence, the last group in the table shows that a linear coupling is allowed in all cases for the piezoelectric-ferroelectric class of transitions. Whenever both phases have inversion symmetry, however, as in the first two groups, then the ion site may or may not possess it. Further, ψ may or may not possess inversion symmetry about the ion site. Since single-ion fluorescence and absorption depend on the local properties of the site, there is no distinction here on the basis of $\underline{q}_\psi$. If the prototypic phase and the order parameter are symmetric under inversion at the ion site, then the first-order perturbation of the electronic levels is given by

$$\langle 0|V_\psi|\mathrm{exc}\rangle \quad ,$$

where V_ψ is a perturbation of the Hamiltonian with the symmetry of ψ. This must be zero for states active in absorption and fluorescence, which must be of odd parity. Here $\langle 0|$ is the ground state and $|\mathrm{exc}\rangle$ is the excited state. Hence we expect these state splittings to show only a higher order dependence on ψ. It should be noted, however, that "forbidden" absorption or fluorescence (e.g., due to thermal fluctuations) is often observed.

While the incommensurate transitions in $TaSe_2$, $BaMnF_4$, and K_2SeO_4 are shown in the table where they would be if they were commensurate, it should be noted that they actually lie in a separate class. Indeed, in the incommensurate phase they do not have a truly periodic structure. The excitations in these materials are fundamentally different and require special consideration [2.17].

From the standpoint of the information available from optical techniques, then, the above categorization in terms of inversion symmetry and $\underline{q}_\psi$ is complete. Thus, we have chosen this as the basis for the selection of examples for this review, as shown in Table 2.1.

In addition to the sections indicated in Table 2.1, we have included separate sections on the application of *nonlinear* optical techniques (Sect.2.9) to the study of phase transitions and on the study of multicritical points (Sect.2.10). Before proceeding to any discussion of results, however, we shall first describe, in Sect. 2.2, the various experimental techniques and the theory behind them.

2.2 Experimental Techniques

In this section we briefly review the techniques involved in various optical measurements. The treatment here is not intended to be exhaustive but rather functional. References to more detailed treatments are given for the interested readers.

2.2.1 Absorption and Fluorescence

We first consider the single-photon processes: absorption and fluorescence. Although these measurements are thus related, there is a wide disparity in the experimental techniques employed, based upon the wavelength region of interest. We shall first consider the techniques used in the infrared (IR), and then those of interest in the visible region of the spectrum.

a) *IR Absorption and Reflection*

In most solid materials the absorption of IR radiation is so strong that a standard transmission absorption measurement is not feasible. The fabrication of suitably thin samples is simply not possible in many cases. However, the same physical information can be obtained from reflection measurements [2.21]. The most intuitive way of seeing this is to recognize that the simple equations relating reflection and refractive indices at nonabsorbing dielectric interfaces hold as well at boundaries of absorbing media. In this case the refractive index (and, hence the dielectric permittivity) become complex quantities. The dielectric constant is then written $\varepsilon = \varepsilon_1 + i\varepsilon_2$ and we see that there are *two* independent components which must be specified as functions of frequency in order to describe the optical properties of the material. In a reflection experiment, these two components determine the amplitude and phase of the reflected light. If the measurement of both amplitude and phase were possible, a single pair of measurements could completely describe the (linear) dielectric response at a given frequency. Until rather recently, however, techniques for this were not developed. In order to appreciate the recent advances, we must consider what was involved previously in the extraction of the components ε_i from a typical reflection spectrum. It is well known that the real and imaginary parts of the dielectric constant are related by the Kramers-Kronig (K-K) relation. The use of this relation, however, requires the knowledge of one part over the entire range of frequencies. This is, of course, not possible experimentally, so approximations are used to extract the imaginary part from the real. These include extrapolation of the low-frequency portion of the data, and assumption of negligible contribution from the high-frequency portion. Both of these assumptions may be unjustified, and can lead to spurious results. The problem is particularly severe in the case of phase transitions where the region of interest lies at the low-frequency end of the spectrum. One is then faced with the problem of relating the form of the spectrum, resulting from the K-K analysis, to the presence of soft modes of vibration at frequencies which range out of the region accessible to IR spectroscopy. There are two problems with this procedure. First, the assumptions required as to the form of the spectral response may not be justified and cannot be checked experimentally. Second, the approximate K-K analysis itself may be grossly in error at the edge of the observable spectral region. The result is often an interpretation of only qualitative validity.

Although efforts have been made to apply this sort of analysis to soft mode spectroscopy near phase transitions, the problems involved have prevented significant progress. Hence, the development of dispersive Fourier transform spectroscopy (DFTS) [2.22] was of major signifcance. The concept of DFTS involves the use of a Michelson interferometer, with the stationary mirror replaced by the reflecting sample. Since this is an interferometric technique, the information on the phase of the reflection is retained. In practice the data are recorded as the movable mirror is scanned through its range, and then the resulting interferogram is Fourier transformed. The information on the phase of the reflection at a given wavelength is contained in the phase of the Fourier transform. Once the complex reflection coefficient $R = r \exp(i\phi)$ is known, then the complex dielectric constant is found from [2.23]

$$\varepsilon_1 + i\varepsilon_2 = [(1 + r \exp(i\phi)/(1 - r \exp(i\phi))]^2 \quad . \qquad (2.6)$$

A complication involved in this procedure is the requirement of having a reference phase to which the reflected phase can be compared. There are various means of accomplishing this important task, such as replacing the sample with the mirror [2.24] or masking part of the sample with a metallic film of known thickness [2.25,26]. The latter techniques offer more reproducible results and have thus become the more popular. The complexities of DFTS have been described in a book by

GRIFFITHS [2.27]. The interested reader is referred there for more detail. The important point, for our purposes, is that the overriding advantage of DFTS is the preservation of the phase information, which is not available in an ordinary reflection measurement, thus eliminating the need for the K-K analysis with its attendant uncertainties.

With the technique of DFTS in hand, the limit of the application of IR techniques is largely determined by the availability of detectors. A review of the available detectors has been provided by KIMMITT [2.28]. IR detectors fall into two classes - thermal detectors, called bolometers, which measure a temperature change, and photoconductive detectors. The application of either type to the region of interest in phase transitions (<100 cm^{-1}) is difficult. However, recent progress has been made in the area of photoconductive detectors, based on very shallow impurity traps in semiconductors. For example, a Ge:Sb detector is capable of a quantum efficiency of about 1% at 10 cm^{-1}. For use at such low energies not only must the detector be cooled to liquid helium temperatures, but so must the entire spectrometer. A detector of this type has been employed, without such extreme cooling, to study IR reflection in KDP and DKDP down to 20 cm^{-1} [2.24].

b) *Optical Absorption and Fluorescence*

By contrast with the situation described above for the case of IR reflection, there has been relatively little progress in apparatus related to one-photon processes in the visible. The Fourier transform technique is naturally applicable in principle to this spectral region as well, but the very short wavelength makes it impractical. Indeed, one could argue that there is little need for improvement. The resolution provided by the typical grating monochromator is already more than sufficient to study the typical fluorescence linewidths in solids, and detectors with good quantum efficiencies are readily available.

Optical absorption has been applied mainly to two phenomena where phase transitions are concerned. These are I) the study of changes in the electronic band gap and II) changes in crystal field splittings of localized electronic levels. In the former case, little quantitative information has been obtained. There exist no microscopic theories to relate the changes observed to structural changes - only phenomenological models [2.29,30].

On the other hand, the study of the effect of the crystalline electric field on ionic electronic levels has yielded important information. This is particularly true in the case of the cooperative Jahn-Teller systems, since in this case the microscopic mechanism of the transition is directly related to the coupling between structural changes and the electronic levels. Information on order parameter behavior, as well as new understanding of symmetry changes, has been obtained. A convenient way of discussing crystalline electric field [CEF] changes is in terms of angular-momentum operators [2.31]. The CEF Hamiltonian is written

$$H_{CEF} = \sum_{n=2,4,6} \sum_{m=-n}^{n} B_n^m x_n O_n^m \quad , \tag{2.7}$$

where the O_{-n}^m are the normalized angular momentum operators, the x_n are reduced matrix elements, and the B_n^m are the CEF coefficients. For a given n, the coefficients B_m^n are often related by symmetry, greatly reducing the number of independent coefficients. From known ionic wave functions it is then possible to calculate the effect of given coefficients on the fluorescence spectrum and to compare the results with experiment. Such analyses have been performed, for $LaAlO_3$ and $PrAlO_3$, as well as for the vanadates [2.7]. Combining this information with that from other experiments has resulted in a very detailed understanding of this class of phase transitions.

2.2.2 Birefringence

In general the term birefringence refers to the difference in refractive index for light of two orthogonal polarizations propagating in a given direction. In any material, a set of coordinates can be chosen such that the refractive index ellipsoid has the form

$$\sum_{i=1}^{3} x_i^2/n_i^2 = 1 \quad , \tag{2.8}$$

where the refractive index for polarization along x_i is n_i. The cartesian coordinates are represented by x_i. For light propagating in a direction $\underline{k}$ the section of the ellipsoid in (2.8) perpendicular to $\underline{k}$ will be an ellipse whose major and minor axes of length $1/n_1$ and $1/n_2$ define the two orthogonal polarization components $\underline{p}_1$ and $\underline{p}_2$. These components travel at the well-defined velocities c/n_1 and c/n_2. The direction $\underline{k}$ is said to be an optic axis of the material if $n_1 = n_2$. For any material of lower than cubic symmetry, there will be at most two such directions. For light traveling along any other direction $\underline{k}$ we can always decompose its polarization into components polarized along $\underline{p}_i$,

$$\underline{E} = (\underline{p}_1 E_1 e^{i\underline{k}_1 \cdot \underline{r}} + \underline{p}_2 E_2 e^{i\underline{k}_2 \cdot \underline{r}}) e^{-i\omega t} \quad , \tag{2.9}$$

where $|\underline{k}_i| = 2\pi n_i/\lambda$, and λ is the wavelength of the light. In a typical birefringence experiment, the light is initially linearly polarized and passes through a birefringent crystal of length ℓ The optical phase difference Γ introduced between the two components is then

$$\Gamma = \ell(\underline{k}_1 - \underline{k}_2) = \frac{2\pi}{\lambda}\ell(n_1 - n_2) \equiv 2\pi\ell\Delta n_{12}/\lambda \quad . \tag{2.10}$$

The accurate analysis of this phase difference is the basic aim of the various birefringence techniques.

At a structural phase transition, some component of the birefringence Δn_{12} will change from zero to a nonzero value. In more generality, if we make the coordinate choice of (2.8) for the protoypic phase we can write, for the distorted phase [2.32]

$$\sum a_{ij}x_ix_j = 1 \quad , \tag{2.11}$$

where

$$a_{ij} \equiv \delta_{ij}n_i^{-2} + \delta a_{ij}$$

and δ_{ij} is the Kronecker delta. It should be noted that $a_{ij} = (\varepsilon^{-1})_{ij}$ where $\varepsilon_{ij} = \alpha_{ij} + \delta_{ij}$ and α is the polarizability tensor. The tensor $\underline{\underline{\delta a}}$ may be expanded in the order parameter as

$$\delta a_{ij} = r_{ij}\psi + s_{ij}\psi^2 + \dots \quad . \tag{2.12}$$

As discussed Sect.2.12, it is possible for symmetry to force the linear term in (2.12) to be zero. In any case, the form of the lowest order coupling may be derived from symmetry considerations. In addition, in cases where domains are possible, the birefringence also affords a very sensitive technique for detection of a phase transition. This has been used, for example, to map out phase diagrams in mixed crystals [2.33], and to investigate the effect of stress on domain structure. However, from the standpoint of the physics of phase transitions themselves, the most important feature is the ability to measure the temperature dependence of ψ near T_c. Since the measurement of Δn is a direct measurement of a *difference* quantity, very high accuracy is possible and the behavior of ψ near T_c may be sensitively determined albeit indirectly by this technique. It should be noted, however, that it involves spatial integration over the optic path through the sample, and may be influenced by surface effects.

a) *Typical Experimental Apparatus*

A typical birefringence experiment is shown in Fig.2.1. Polarized light is incident on the sample (S). After traversing the sample it is, in general, elliptically polarized. The function of the compensator (C) is to restore linear polarization and, in so doing, to measure the phase difference of the two polarization components. When the axes of the compensator are aligned in the directions of $\underline{p}_1$ and $\underline{p}_2$ [2.9] and the output polarization is linear we have

$$\Gamma_c = \frac{2\pi}{\lambda}\,\ell(\Delta n_{12}) \pm \pi m \quad ,$$

where Γ_c is the relative retardation introduced by the compensator and m is an integer. The existence of this condition is detected by the final polarization analyzer. The compensator-analyzer combination are adjusted for optimum rejection of the transmitted beam.

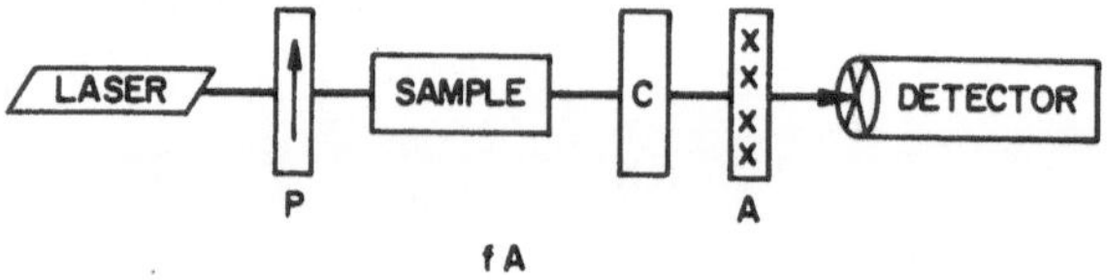

Fig. 2.1. Schematic of a typical birefringence experiment

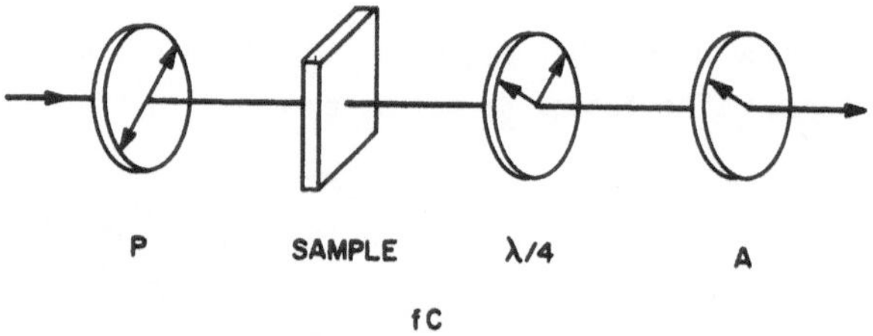

Fig. 2.2. Senarmont method for compensation in a birefringence experiment

Various compensator designs are used [Ref.2.34, p.559]. Several depend on the use of uniaxial crystals [$n_1 = n_2 \neq n_3$ in (2.8)] with a variable path length. These include the Babinet, Soleil, and Berek compensators. A different technique for compensation is shown in Fig.2.2. This is known as the Senarmont method [2.35]. In this technique the incident light is polarized at 45° to the major and minor axes of the cross section of the index ellipsoid. A quarter-wave plate follows the sample and is aligned with the *incident* polarization. Under these conditions simple algebra shows that the emergent light is linearly polarized with the polarization rotated by an amount equal to half the retardation introduced by the sample. The angle of this rotation is then an easily measured quantity. This method obviously is applicable only when the axes $\underline{p}_1$ and $\underline{p}_2$ are fixed and known - but this is precisely the case in a structural phase transition. Thus the Senarmont method is one of those most commonly used to obtain order parameter information.

Variations of these basic techniques include polarization modulation techniques [2.35-37] and computer analysis of data as a function of Γ_c [2.6]. By such techniques a resolution of better than 10^{-7} in the absolute value of Δn may be achieved [2.38].

An interesting variation of the Senarmont method with a sampling technique has been developed for ultrafast response [2.39], in the nanosecond regime (τ_r = 6 ns). This technique has been applied, for example, to the studies of time-resolved electrooptic response in $BaTiO_3$ and of optical damage in $LiNbO_3$ [2.40].

2.2.3 Light Scattering

A natural division of the field of light scattering may be made on the basis of the frequency of the excitations ($\Delta\omega$) under study. The various experimental techniques operate in frequency regimes which overlap only slightly. There are three basic experimental techniques used in light scattering: grating spectroscopy (100 THz > $\Delta\omega$ > 3 GHz), interferometry (50 GHz > $\Delta\omega$ > 1 MHz), and correlation spectroscopy (10 MHz > $\Delta\omega$). (Note that 1 THz is the same frequency unit as 33.3 cm^{-1} and 4.16 meV). We shall discuss only the first two of these techniques here, as the latter has seen little application to solids.

a) *General Background*

The geometry of a typical light scattering experiment is shown in Fig.2.3. The incident optical field is E_1, of frequency ω_1, wave vector $\underline{k}_1$, and linear polarization $\underline{\varepsilon}_1$. It traverses the scattering volume, which can be described by a space- and time-dependent polarizability tensor $\delta\alpha_{ij}(r,t)$. The scattered light is observed at a scattering angle of θ at the field point $\underline{r}$, with frequency ω_2, wave vector $\underline{k}_2$, and polarization $\underline{\varepsilon}_2$. For a perfectly homogeneous polarizability, the reradiated (scattered) field would sum to zero for all but the exact forward direction. Thus, the average value of α is related to the refractive index by $\alpha = n^2 - 1$. For a perfectly static polarizability, the scattered light field would have the same frequency as the incident field. Experimentally one measures the strength, polarization, frequency, and wave vector of the scattered field. These quantities carry information on the strength, symmetry, temporal and spatial properties, respectively, of $\delta\alpha_{ij}$. The information revealed by the scattering experiment is contained [2.41] in the space-time correlation function of the polarizability tensor elements

$$\langle\delta\alpha_{ij}(\underline{r},t)\delta\alpha^*_{k\ell}(\underline{r}=0,\ t=0)\rangle \quad . \tag{2.13}$$

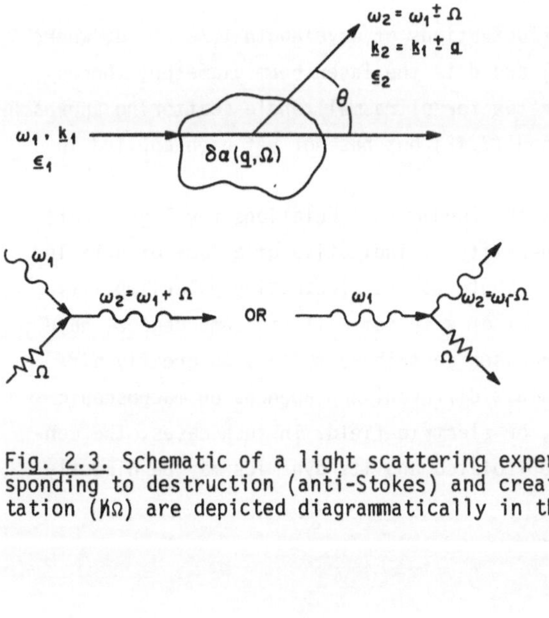

Fig. 2.3. Schematic of a light scattering experiment. Scattering processes corresponding to destruction (anti-Stokes) and creation (Stokes) of an elementary excitation ($\hbar\Omega$) are depicted diagrammatically in the lower part of the figure

The scattered field is proportional to the $\underline{q}^{th}$ spatial Fourier component of $\delta\alpha_{ij}$, where $\underline{q} = \underline{k}_2' - \underline{k}_1'$ and $k_i' = nk_i$. At $\omega_2 = \omega_1 \pm \Omega$, the Ωth Fourier component of $\delta\alpha$, $\delta\alpha(\underline{q},\Omega)$, is observed. The various polarization components of the scattered field amplitude are proportional to

$$E_2^{(i)} \approx (\text{const})k_1^2E_1^{(j)}\delta\alpha_{ij}(\underline{q},\Omega) \quad . \tag{2.14}$$

So the scattered intensity of polarization i is just proportional to

$$I_i \propto k_1^4E_1^{(j)}E_1^{(m)} \langle\delta\alpha_{ij}(q,\Omega)\delta\alpha_{im}(\underline{q},\Omega)\rangle \quad , \tag{2.15}$$

where the summation of the repeated indices j and m is implied.

Implicit in this expression are the kinematics of the scattering process, which correspond to overall momentum and energy conservation

$$\hbar\underline{k}_2 = \hbar\underline{k}_1 \pm \hbar\underline{q} \quad \text{and} \quad \hbar\omega_2 = \hbar\omega_1 \pm \hbar\Omega \quad . \tag{2.16}$$

Hence $\underline{q}$ is called the scattering momentum transfer. The - and + signs correspond to gain and loss, respectively, of a quantum of energy $\hbar\Omega$ by the medium, and are called the Stokes and anti-Stokes components of the spectrum. For a system in thermal equilibrium the spectrum is symmetric about ω_1 when $\hbar\Omega \ll kT$, but more generally exhibits the relative demagnification by $\exp(-\hbar\Omega/kT)$ on the anti-Stokes side required by the principle of detailed balance. From Fig.2.3 we see that

$$\hbar q = 2\hbar k_1 \sin(\theta/2) \quad . \tag{2.17}$$

Thus variation of θ permits probing of fluctuations of wavelength $\lambda_1/2 < \lambda < d$, where λ_1 is the laser wavelength ($\approx 10^{-5}$ cm^{-1}) and d is the laser beam diameter, thereby defining the diffraction limit. An apparatus for ultrasmall angle scattering approaching this limit has recently been developed [2.42] but has not yet been applied in the study of phase transitions.

Equations (2.15) and (2.16) represent the fundamental relations for light scattering experiments. However, their very generality is indicative of a lack of detailed physical content. Many processes may contribute to the fluctuating polarizability $\delta\alpha_{ij}$. In fact, virtually any dynamic aspect of a system will, to some degree, have an influence. Fortunately, different processes contribute to $\delta\alpha_{ij}$ on greatly different time or length scales and may exhibit different dependences on macroscopic variables such as temperature, pressure, or electric field. In such cases, the contribution of a particular process may be isolated and its dynamics may be directly measured by light scattering.

$\hbar = h/2\pi$ (normalized Planck's constant)

b) *Scattering from Coupled Modes*

In crystalline solids the elementary excitations (phonons) are often reasonably well described within the quasi-harmonic approximation. However, when excitations are coupled, the situation becomes more complicated. A concise formulation of this problem has been given by FLEURY [2.43] based upon a generalized susceptibility. Let us consider, for example, the case where an acoustic mode couples to the soft mode at a structural phase transition. Such a coupling will exist in general if the latter is Raman active. For both modes we can define a generalized susceptibility as a function of wave vector $\underline{q}$ and frequency ω. This is the Fourier transform of the susceptibility $\chi_{\psi,X}(\underline{r},t)$ of the crystal to a field which couples linearly to ψ (for the soft mode) or to the strain X (for the acoustic mode). For example, for the soft mode

$$\chi_\psi(\underline{r},t) = \left(\frac{i}{\hbar}\right) \langle\psi(\underline{r},t)\psi(0,0)\rangle \tag{2.18}$$

and the Fourier transform is given by (2.4). The scattering spectrum for the *uncoupled* modes is then given by

$$S_i(\underline{q},\omega) = -[n(\omega) + 1]\,\mathrm{Im}\{\chi_i(\underline{q},\omega)\} \quad , \tag{2.19}$$

where $[n(\omega)+1] = [1-\exp(-\beta\hbar\omega)]^{-1}$ and a constant geometric factor is omitted. In the simplest case we use the quasi-harmonic approximation, where

$$\chi_h(\underline{q},\omega) = \mathrm{const}(\omega_{\underline{q}}^2 - \omega^2 - 2i\Gamma_{\underline{q}}\omega)^{-1} \quad .$$

The spectrum, from (2.19), thus becomes

$$S_h(\underline{q},\omega) = [n(\omega) + 1]\frac{2\Gamma_{\underline{q}}\omega}{(\omega_{\underline{q}}^2-\omega^2)^2+4\omega^2\Gamma_{\underline{q}}^2} \quad . \tag{2.20}$$

When $\omega_{\underline{q}} \gg \Gamma_{\underline{q}}$, the two poles of this expression are well approximated by a pair of Lorentzians at $\omega = \pm\,\omega_{\underline{q}}$ of half-width at half maximum (HWHM) $\Gamma_{\underline{q}}$. However, when $\omega_{\underline{q}} < \Gamma_{\underline{q}}$ the two peaks of S_h coalesce into one and in the limit $\omega_{\underline{q}} \ll \Gamma_{\underline{q}}$ there is only one physically relevant frequency, the width of the single remaining Lorentzian, centered at zero, which is given by $\omega_{\underline{q}}^2/2\Gamma_{\underline{q}}$ (HWHM).

So far we have not discussed the intensity of the scattering. This matter becomes very important when considering coupled modes. For the two modes we are considering, the total susceptibility is given by [2.43]

$$\chi_T(\underline{q},\omega) = \sum_{i,j=a,\psi} F_iF_j\chi_{ij}(\underline{q},\omega) \quad , \tag{2.21}$$

where a and ψ denote the acoustic and soft modes, respectively, and the generalized susceptibilities χ_{ij} for the coupled modes are given by

$$\chi_{ii} = \frac{\chi_i}{1-A^2\chi_a\chi_\psi} \; ; \quad \chi_{a\psi} = \frac{A\chi_a\chi_\psi}{1-A^2\chi_a\chi_\psi} \quad . \tag{2.22}$$

Here A is a coupling parameter representing the strength of the interaction of the modes and the F_i's represent the coupling of the uncoupled modes to the experimental probe, in this case, light. This very important result has sometimes been overlooked. From (2.21) it is possible to understand the difference in spectra observed in the same system by various techniques, such as neutron scattering and light scattering. The values of F_ψ/F_a for the different scattering processes may be widely different. This is particularly true near a phase transition where the soft mode is not Raman active above T_c. In that case the strength of the coupling F_ψ is proportional to ψ below T_c, while for neutron scattering there is no corresponding selection rule effect. Thus, the relative intensity of the soft and acoustic mode spectra is vastly different in the two techniques. The shape of the total spectrum, $S_T(\underline{q},\omega) \approx \mathrm{Im}\{\chi_T(\underline{q},\omega)\}$, determined by the interference effects in (2.21), may be significantly different.

c) *Critical Exponents and Soft Mode Spectra*

The extension of the above static critical exponents to dynamic phenomena requires some discussion. The fluctuations of ψ are known as the *soft mode* of the phase transition. However, the extraction of "critical exponents" from soft mode spectra has often not been treated adequately. In addition to the obvious experimental difficulties associated with analysis of multicomponent spectra, particularly in the presence of mode coupling and background effects, there are some potential theoretical pitfalls of which one must be wary. In the simplest case where mode coupling does not affect the order parameter dynamics near T_c, and where the quasi-harmonic approximation [2.20] is valid, the relationship between the soft mode spectrum and the critical exponents is clear. Specifically in that case one uses the fluctuation-dissipation theorem to define a frequency parameter ω_s through the relation

$$\int d\omega \, \frac{\chi''_\psi(\omega)}{\omega} = \chi_\psi(\omega = 0) = 1/\omega_s^2 \quad . \tag{2.23}$$

For the quasi-harmonic form [2.20] of $\chi_\psi(\omega)$ this definition identifies ω_s with ω_q.

In practice, mode coupling leads to more complicated forms of $\chi_\psi(\omega)$ and it is useful to define a characteristic frequency and a dynamic critical exponent, z. The present understanding of critical dynamics has recently been reviewed by HOHENBERG and HALPERIN [2.44]. We extract a few of their remarks as relevant here.

They describe two ways of defining the characteristic frequency ω_ψ. Equation (2.23) is, in essence, a statement relevant only to static exponents. A more general definition involving the "median frequency" $\omega_\psi(k)$ is implicitly given by

$$\int_{-\omega_\psi}^{\omega_\psi} \frac{d\omega}{2\pi} S(\underline{q},\omega) \equiv \tfrac{1}{2} S(\underline{q}) \quad . \tag{2.24}$$

For an overdamped Lorentzian spectrum centered at $\omega = 0$ this definition differs considerably from the $\omega_{\underline{q}}$ defined in (2.23). It is only in the underdamped case that $\omega_\psi = \omega_{\underline{q}}$.

The critical behavior of the characteristic frequency depends upon the universality class to which the transition belongs. Even when z can be expressed in terms of the static exponents, the relationship may not be the same for different universality classes. For example, in the case of many structural transitions the order parameter is coupled to an auxiliary conserved density (like the energy density). In this case

$$\omega_\psi(\underline{q} = 0) \propto \xi^{-z} \propto t^{+\nu z} \quad , \tag{2.25}$$

where ξ is the correlation length and ν its static critical exponent, and scaling arguments yield the relation $z = 2 + \alpha/\nu$. For example, the strontium titanate transition is expected to exhibit α to be very small and $\nu \approx 2/3$, according to the renormalization group analysis described in Sect.2.10. Thus, $\omega_\psi \propto t^{1.33}$ is the approximate behavior expected. Note that in this case $\omega_\psi \sim t^{4\beta}$. As discussed in Sect.2.6.2 the relationship between the actual observed spectra and the characteristic frequency ω_ψ, in that case, is complicated by selection rules, mode couplings, the nonharmonic form of χ_ψ, etc., so that no definitive comparison of theory and experiment regarding z has yet been made. In mean field theory this definition of z is not useful because dynamic scaling may break down.

d) *Intensity of Scattering*

In addition, the selection rule effect alluded to above can provide a very sensitive test for deviations from mean field theory, as we shall now show for the case of uncoupled modes. The total integrated intensity of the spectrum in (2.20) behaves as $(1/\omega)\,\mathrm{Re}\{\chi_{\underline{q}}(\underline{q},\omega = 0)\} = \omega_{\underline{q}}^{-2}$. In cases where the susceptibility is more complex than that which leads to (2.20), this relationship (generally known as the fluctuation-dissipation theorem) may serve as a static definition of $\omega_{\underline{q}}$ as discussed in Sect. 2.1.1. As T_c is approached from below this diverges. However, if the mode is not Raman active above T_c, the behavior of F_ψ makes the total intensity, within MFT, behave as

$$I_{TOT} \approx \psi_0^2 \omega_{\underline{q}}^{-2} \tag{2.26}$$

which is constant near T_c in MFT. Another way of seeing this is to say that when the linear term in (2.12) is zero, the total intensity, the integral of (2.15), will be proportional to the spatial Fourier transform of $G_2(\underline{r},0) = \langle\psi^2(\underline{r},0)\psi^2(0,0)\rangle$ at wave vector $\underline{q}$. If $\psi = \psi_0 + \delta\psi$ and $\delta\psi$ is small, this reduces to (2.26) when we recognize the proportionality [2.43] of $1/\omega_q^2$ to $\langle\delta\psi(\underline{r})\delta\psi(0)\rangle$. Thus, the MFT result of (2.26) amounts to dropping the term $\langle[\delta\psi(\underline{r})]^2[\delta\psi(0)]^2\rangle$. When this term becomes important we then expect an increase in the intensity as T_c is approached. It should be noted that such an increase may be much more reliably determined than a deviation from a strong power law divergence. Hence light scattering can be a sensitive technique for detection of deviations from MFT in appropriate cases.

More accurately, it should be pointed out that the correlation function $G_2(\underline{q} = 0, \omega = 0)$ appears also in the expression for the critical component of the specific heat in the case when ψ^2 represents an energy density. Hence the temperature dependence of the integrated light scattering intensity should be that of the critical part of the specific heat. Indeed the MFT prediction of a step in I_T at T_c represents the same thing as the MFT prediction of a step in C_p. The advantage in light scattering is that there is often no strong background superimposed on the weak anomaly as there usually is in the case of the specific heat.

Equation (2.21) holds more generally for the case of n pairwise coupled excitations, in which case the determination of χ_{ij} requires diagonalization of an $n \times n$ matrix and the knowledge of $n(n-1)/2$ complex coupling coefficients.

With this background of the theory of light scattering, we now turn to a discussion of the two main techniques used in solids.

e) *Grating Spectroscopy*

The term "Raman scattering" initially referred to the study of vibrations in molecules and, later, optic vibrational modes in solids. Classically, the interpretation of vibrational Raman scattering is simply that a vibration of appropriate symmetry may modulate the polarizability tensor, hence introducing into the scattered light components at $\omega_L \pm \Delta\omega$ where ω_L is the laser frequency and $\Delta\omega$ is that of the vibration. The energy range of interest in Raman scattering typically extends from $\approx 1\ cm^{-1}$ (0.03 THz) out to $\approx 3000\ cm^{-1}$ (100 THz) and therefore is usually studied by grating spectroscopy. The typical frequency unit used is the wave number, cm^{-1}, so we shall employ that unit in discussing Raman results. The reader should note that when an angular frequency (ν) is quoted in Hz, or other equivalent units, a factor of 2π is understood, as $\omega(s^{-1}) \equiv 2\pi\nu(Hz)$.

The contribution of a particular excitation to the scattered spectrum is contained in the correlation function of (2.15). Many detailed descriptions of the Raman scattering process have appeared and we direct the reader to these [e.g.2.45]. For our purposes here a simple classical approach will suffice. If we let u represent the value of the vibration coordinate for the excitation in question, then there may be a linear term in $\delta\underline{\alpha}$, $\delta\alpha_{ij} = A_{ij}u$. This leads to a relation of the scattered field components $E_2^{(i)}$ to the incident field $E_1^{(j)}$, namely $E_2^{(i)} = R_{ij}E_1^{(j)}$, as in (2.14). The tensor $\underline{R}$ is called the Raman tensor. The nonzero components of R_{ij} for a given excitation are determined by symmetry, but their magnitudes are determined by microscopic details of the system. If a vibrational mode is of a symmetry that allows a nonzero value for any component of $\underline{R}$, that mode is said to be *Raman active*. By appropriate analysis of the spectrum in terms of the incident and scattered polarization it is possible to investigate all six independent components of the Raman tensor and thus to classify the symmetry of the observed excitations. Often only a few components of the Raman tensor will be nonzero for a given excitation, by symmetry, thus simplifying the task of classification.

It has been shown [2.46,47] that at least one component of the soft mode of a continuous structural phase transition is always Raman active below T_c. For this reason Raman scattering is an important technique of general usefulness in the study of phase transitions. In the case of displacive structural phase transitions, the soft optic mode typically has an energy on the order of 100 cm^{-1} well below T_c and exhibits a strong temperature dependence as T_c is approached. As stated above, the usual experimental apparatus is a grating double monochromator, that is, two Czerny-Turner spectrometers operated in tandem. Typical resolution of such an instrument is 0.2 cm^{-1} while it is capable of rejecting stray laser light by a contrast factor of up to 10^{-8} for $\Delta\omega > 5$ cm^{-1}.

f) *Interferometry*

In contrast to Raman scattering, the term Brillouin spectroscopy usually refers to the study of *acoustic* excitations. At the wave vectors typical of light scattering, the frequencies of these excitations in normal solids lie in the range 10-60 GHz (0.3 to 2 cm^{-1}). Thus the experimental technique in this case is typically Fabry-Perot interferometry. This technique, based upon interference of light multiply reflected between mirrors, is described in the literature [Ref.2.34, pp.323-341]; hence we shall not derive it here. The result of this analysis is to show (for a pair of plane parallel mirrors) that the transmission function is periodic in $1/\lambda$ with a period of $2Ln/c$, where L is the distance between the mirrors and c is the speed of light. This period is referred to as the free spectral range of the instrument. The instrument may be scanned either by physically changing L or by varying the interplate refractive index n. The resolution Γ_{inst} is governed largely by the quality of the mirrors and the spacing, and is usually discussed in terms of the "finesse", $F = (2Ln/c)/\Gamma_{inst}$. Typical maximum values of F are in the range 50 to 75. It is important to keep the periodic nature of the Fabry-Perot transmission function in mind when considering experimental data, because scattering in the various "orders" may overlap, resulting in a spectrum confusing in appearance. The contrast of a conventional single-pass Fabry-Perot ($\sim 4F^2/\pi^2$, typically $> 10^3$) is not as good as a grating monochromator, but its resolution capability is much better. Typically Γ_{inst} = 1.0 GHz is used for the study of crystals, but resolutions as fine as 10 MHz are possible.

The analysis of the scattering process is actually identical to that for Raman scattering, except that a different class of excitations is usually involved. An important difference, characteristic of the excitations, is that the acoustic phonons are strongly wave vector dependent. The excitation frequency is $\Delta\omega = \underline{q} \cdot \underline{v}$, where v is the sound velocity. In contrast, the optical phonon frequencies usually studied by Raman spectroscopy typically have zero dispersion at the zone center. Hence, by virtue of (2.17), Brillouin spectra typically display stronger angular dependence than do Raman spectra.

From the standpoint of the study of structural phase transitions, there are three situations in which Brillouin scattering is of interest. The most obvious one is the ferroelastic transition, in which the order parameter is strain itself and, hence, the soft mode is an acoustic phonon at $\underline{q}$ = 0. In this case, Brillouin scattering provides a means for direct study of the soft mode. A second application lies in the study of interactions between the soft mode and strain. This is reflected in movement and/or broadening of Brillouin components of appropriate symmetry. The third situation, potentially the most interesting for the understanding of phase transitions, is the case in which it possible to approach a phase transition closely enough that the soft optic mode moves into the range of Brillouin scattering. Not only does this enhance the interactions with acoustic phonons, but also it is precisely in this range that deviations from quasi-harmonic behavior may be most readily observed. This has now been done in at least three materials (see Sects.2.4,2.5.3,2.6.2).

g) *Recent Advances of Technique*

In any light scattering experiment there are two components in the scattered light: elastic and inelastic. The elastic component results from scattering from static inhomogeneities in the crystal. In most cases the interesting information is contained in the inelastic component, which is often orders of magnitude weaker than the elastic. Hence it is important 1) to enhance the signal gathering capabilities for the inelastic component as much as possible and 2) to reject the stray elastic light. With one exception, the improvements in technique we shall mention fall into the area of rejection of stray light. That exception is the advent of holographic gratings which greatly improve the contrast and resolution of grating monochromators [2.48,49].

The stray light rejection techniques fall into two categories: resonant reabsorption and compound interferometry. The resonant reabsorption technique is based on the overlap of a strong narrow absorption in molecular I_2 vapor by the gain curve of an Ar^+ laser operated at 5145 A. This technique is applicable to both Raman and Brillouin scattering. A typical apparatus for use in Brillouin scattering is shown schematically in Fig.2.4. The laser, operated single mode, is tuned on the I_2 absorption by an intracavity etalon, and is monitored by use of an external vapor cell and a spherical Fabry-Perot (SFP). Under these conditions it is possible to attenuate the stray elastic component by as much as 10^{+7} while light at frequencies as close as 0.5 GHz is attenuated by less than one order of magnitude. A disadvantage of this technique, especially when used for high-resolution work, such as in Brillouin scattering [2.41,50] is that other absorptions in the I_2 vapor introduce structure in the instrumental response which can mask features of interest in a spectrum. It has proved possible, however, to remove this structure by computer processing of the signal and thereby to recover quantitatively the spectral profile of the inelastic component [2.50]. Results discussed in Sect.2.4 are based in part upon this technique.

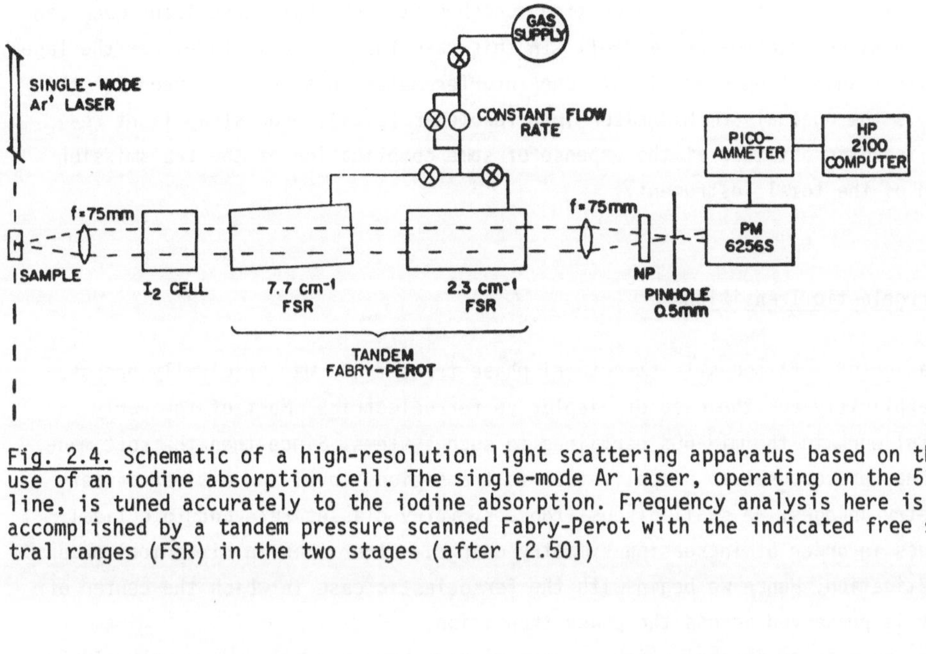

Fig. 2.4. Schematic of a high-resolution light scattering apparatus based on the use of an iodine absorption cell. The single-mode Ar laser, operating on the 5145A line, is tuned accurately to the iodine absorption. Frequency analysis here is accomplished by a tandem pressure scanned Fabry-Perot with the indicated free spectral ranges (FSR) in the two stages (after [2.50])

An important ingredient in the success of this technique is the associated use of a compound planar Fabry-Perot. The problem in any compound interferometer is one of preventing interferometric interaction among the separate stages. One way of doing this, developed by CANNELL and BENEDEK [2.51] is to use two spherical Fabry-Perot interferometers in tandem. This method is capable of the very high resolution characteristic of a spherical Fabry-Perot (≈ 15 MHz or better). A second method, reported by LYONS and FLEURY [2.50], is to use one planar Fabry-Perot off axis, tuned to the second by a slight tilt. This combination is better adapted to the resolution required for study of spectra at structural phase transitions, on the order of 1 GHz, and is free from the disadvantage of a fixed free spectral range inherent in the spherical Fabry-Perot. In either case, the free spectral ranges of the two stages are chosen such that significant overlap of the individual transmission functions does not occur in the first several orders - thus greatly multiplying the effective finesse. A qualitatively different compound interferometer is the multipass Fabry-Perot [2.52] in which the light traverses the *same* interferometer a number of times. This results in a smaller improvement in finesse, but a very large improvement in contrast. Thus, the various compound interferometric techniques represent the answers to different problems and the choice among them hinges on the resolution required and the presence of interfering spectral components which may cause order-overlap problems in a multipass instrument. Very recently SANDERCOCK [2.53] has reported the successful development of a tandem multipass interferometer, which is thus capable of both high contrast and large free spectral range.

It should be mentioned that it is possible to combine an interferometric technique for stray light rejection with a grating monochromator so that these techniques can be used in Raman spectrosocpy as well. In this case the desire is to *reject* the laser line while transmitting other light. One interferometer that has been used for this purpose is the unequal-arm Michelson interferometer [2.54]. Good stray light rejection is again obtained at the expense of some complication of the transmission function of the total instrument.

2.3 Ferroelastic Transitions

The relation of soft modes to structural phase transitions was originally proposed [2.55] explicitly for the case of displacive ferroelectrics. Most of the early experimental work in the mid-60s pertained to such systems. Since then the soft mode concept has proved useful in application to transitions involving a wider variety of symmetry changes. We shall discuss those symmetry classes relevant to optical techniques in order of increasing symmetry change, rather than in historical order of investigation. Hence we begin with the ferroelastic case in which the center of symmetry is preserved across the phase transition.

The properties of ferroelastics in general have been recently reviewed by TOLEDANO [2.56]. Ferroelastic materials are characterized by the existence of a spontaneous strain, analogous to the spontaneous polarization of a ferroelectric. The terminology which has evolved to classify such materials [2.57,58] may be confusing at times. In this section we shall consider the simple case of a pure (nonferroelectric) ferroelastic. As mentioned in Sect.2.1, the symmetry properties of pure ferroelastic transitions are different from ferroelectrics. In particular, insofar as light scattering is concerned, a soft mode related to a ferroelastic transition will be Raman active on both sides of the transition. Thus, these materials are particularly interesting in that they afford an opportunity to study the soft mode by light scattering *through* the phase transition, rather than only below. Ferroelastic transitions of the cooperative Jahn-Teller (CJT) type are of particular interest. CJT materials represent one of only a few classes of phase transitions where a microscopic understanding of the transition mechanism has been achieved. They are also amenable to study by varied optical techniques. We shall begin this section with a brief listing of their properties.

2.3.1 Cooperative Jahn-Teller Systems

A cooperative Jahn-Teller transition is driven by an electron-phonon interaction. The theory of such systems was first worked out by ELIOTT et al. [2.59], and later supplemented by PYTTE [2.60]. On an intuitive level, one can understand the phase transition by simply considering the fact that if a lattice distortion splits a degenerate electronic state, then the accompanying decrease in electronic energy, if the temperature is low enough to depopulate the upper state, may overcome the increase in lattice energy caused by the distortion. The distortion hence becomes energetically favorable and takes place.

The intuitive idea expressed above has been put in more rigorous terms [2.7,59,60]. The simplest situation is that of the interaction of a crystal field doublet with a single nondegenerate phonon mode. In this case, the relevant Hamiltonion becomes [2.7]

$$\mathcal{H} = \sum_{\underline{q}} \hbar\omega(\underline{q})(a_{\underline{q}}^{\dagger}a_{\underline{q}} + \tfrac{1}{2}) + \sum_{\underline{q},n} \exp(i\underline{q}\cdot\underline{R}_n)\xi(\underline{q})S^z(n)(a_{\underline{q}} + a_{-\underline{q}}^{\dagger}) \quad . \tag{2.27}$$

Here $a_{\underline{q}}$ is the creation operator for the phonon of wave vector $\underline{q}$, $\underline{R}_n$ is the position of the nth ion, whose electronic state is specified by the pseudospin operator $S^z(n)$. The coupling coefficients $\xi(\underline{q})$ between the pseudospin and the vibrational states should not be confused with the correlation length ξ defined earlier. The form of this equation can be greatly simplified by introducing the displaced "mixed mode" operator

$$\gamma_{\underline{q}}^{\dagger} = a_{\underline{q}}^{\dagger} + \frac{\xi(\underline{q})}{\hbar\omega(\underline{q})} S_{\underline{q}}^z \quad . \tag{2.28}$$

With this substitution (2.7) becomes

$$\mathcal{H} = \sum_{\underline{q}} \hbar\omega(\underline{q})(\gamma_{\underline{q}}^{\dagger}\gamma_{\underline{q}} + \tfrac{1}{2}) - \sum_{\underline{q}} \frac{|\xi(\underline{q})|^2}{\hbar\omega(\underline{q})} S_{\underline{q}}^z S_{-\underline{q}}^z \quad . \tag{2.29}$$

The pseudospin part of (2.29) is just the Ising Hamiltonian which possesses well-known solutions, and is known to undergo a second-order phase transition. More complicated versions of the Hamiltonian have been treated in the literature, and the reader is referred to the review by GEHRING and GEHRING [2.7] for details.

It should be clear that Jahn-Teller systems are ideally suited to study by optical techniques. By their very nature, they involve two phenomena, both of which are independently observable by appropriate optical experiments. The ionic electronic states may be studied by absorption and fluorescence, as well as by electronic Raman scattering. The phonons coupled to those states may be observed by light scattering, either Raman or Brillouin, depending on the nature of the transition. Moreover, as mentioned above, the relevant phonons, in the case of ferroelastic transitions, are Raman active on both sides of T_c.

2.3.2 $PrAlO_3$: A Prototypical CJT System

Of the many CJT systems known and investigated we have chosen to present in detail here results in $PrAlO_3$. Similar results not discussed here have been obtained in other rare earth compounds [2.7]. $PrAlO_3$ exhibits a sequence of four phase transitions, three of which are ferroelastic and may be interpreted in terms of CJT effects. A convenient way of viewing the sequential phase transitions in $PrAlO_3$ is in terms of the staggered rotations of the AlO_6^{3-} octahedra. As shown in Fig.2.5, in the three lower temperature phases the unit cell is doubled from the cubic in the (111) direction because the rotation is staggered, in opposite senses in adjacent unit cells. The axis of that rotation defines the various phases, while its magnitude remains relatively constant, except in the high-temperature, cubic perovskite, phase, where it vanishes. On reducing the temperature from that phase, which is stable above 1640 K, a zone boundary transition to a trigonal phase takes place, in which the staggered rotation axis is along (111). This structure is favored by lattice forces alone. This transition, which also occurs in $LaAlO_3$, is thus very similar to that which occurs in $SrTiO_3$. It is followed by a first-order transition

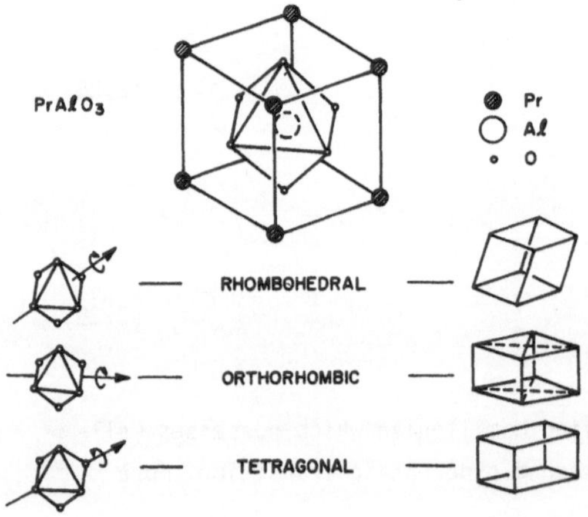

Fig. 2.5. Cubic perovskite unit cell, together with the three possible daughter phases: rhombohedral (trigonal), orthorhombic, and tetragonal, resulting from antiphased AlO_6 octahedral rotations about the indicated directions in adjacent cubic unit cells

to an orthorhombic structure at 210 K, in which the octahedral rotation is about (101). Subsequently a transition to monoclinic symmetry occurs at 151 K as the axis begins to swing continuously toward the (001) direction, where the structure would become asymptotically tetragonal. Before that happens, however, a fourth transition takes place at 118 K which may be viewed as a slight movement of the rotation axis out of the [101] plane.

The electronic levels of the Pr^{3+} ions have been studied by fluorescence [2.61], Raman scattering [2.62,63], and optical absorption [2.64]. The results of the fluorescence and Raman studies are summarized in Fig. 2.6, along with the results of two model calculations for the effects of the crystalline electric field. The data, shown in Fig.2.6b, clearly are in disagreement with the calculation in Fig.2.6a, which was performed as mentioned above [2.6] taking into account only first-order perturbation terms within the $J=4$ manifold of the Pr^{3+} ion. The calculation in Fig.2.6c, which includes J mixing with the $J=5$ manifold, shows satisfactory agreement.

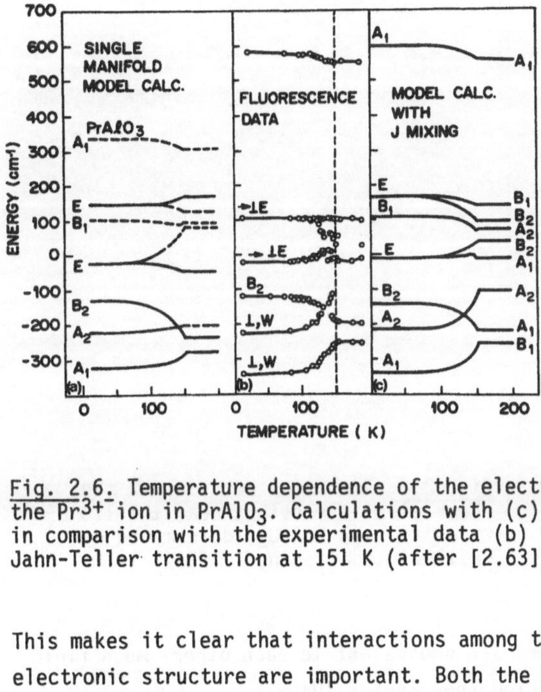

Fig. 2.6. Temperature dependence of the electronic levels of the $J=4$ manifold of the Pr^{3+} ion in $PrAlO_3$. Calculations with (c) and without (a) J mixing are shown in comparison with the experimental data (b) in the vicinity of the cooperative Jahn-Teller transition at 151 K (after [2.63])

This makes it clear that interactions among the different J manifolds of the ionic electronic structure are important. Both the Raman and fluorescence spectra provide information on the splitting of the ground state of the manifold, which in this case is an order parameter for the transition. Similar information is obtained from absorption [2.64]. The latter was found to be in agreement with the Raman and fluorescence measurements, as well as with ESR and structural data. This comparison is presented in Fig.2.7, where the splitting of the ground doublet is shown plotted with the rotation direction of the octahedral axis and also with the macrosopic

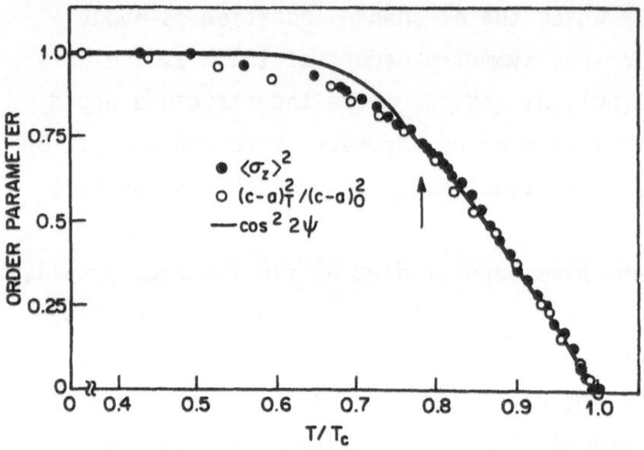

Fig. 2.7. Temperature dependence of the order parameter below the 151 K CJT transition in $PrAlO_3$, determined from optical absorption (●●●), strain (ooo), and EPR (——) measurements (after [2.64])

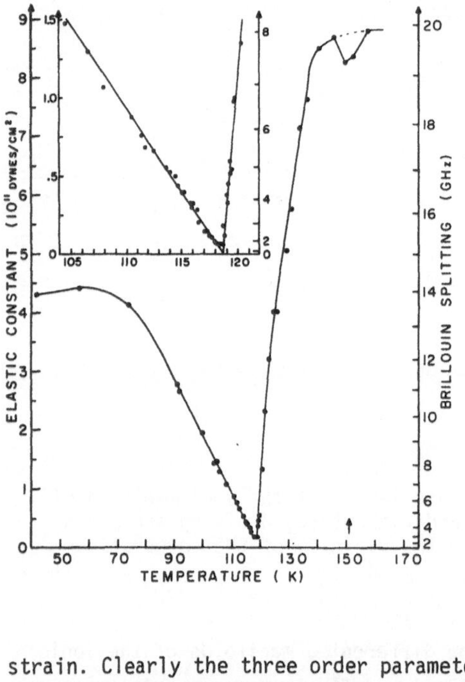

Fig. 2.8. Anomalous behavior of the shear elastic constant near the 118 K transition in $PrAlO_3$, determined by Brillouin scattering [2.65]

strain. Clearly the three order parameters are equivalent to each other. Mean field behavior is observed throughout, as expected (see Sect.2.10).

Thus far we have said nothing about the 118 K transition. In fact it causes little observable change in the electronic levels, except for the small dip in the splitting shown in Fig.2.7. It is reflected, however, in the phonon spectra of the material, as evidenced weakly in the Raman spectrum of the external modes [2.62] and quite strongly in the Brillouin spectrum [2.65]. In fact, this transition was first detected by the presence of the anomaly in the position of a Brillouin line, displayed in Fig.2.8. This situation is particularly interesting because not only does the mode soften by fully 99%, but it is observable on both sides of T_c due to the ferroelastic character of the transition. Curiously this transition shows very small effects when probed with all other experimental techniques, indicating that it may be a member of a very restricted class of pure strain transitions [2.65]. The mechanism of the transition has however been recently ascribed to a coupling between an acoustic mode and a temperature-dependent optic mode [2.60]. In turn the temperature dependence of the optic mode is due to the CJT interaction. Hence this transition may be related to the others albeit in a rather indirect way.

From the above discussion it is clear that in the case of $PrAlO_3$ a very complete and unified picture of the phase transitions in this material has resulted from the application of a variety of optical techniques, capable of probing all the various aspects of the CJT interaction.

2.4 Nonpiezoelectric Ferroelectric: $Pb_5Ge_3O_{11}$

In the nonpiezoelectric ferroelectric materials, the prototypic phase possesses a center of symmetry which is lost at the transition. In most such materials, there is no optical technique that directly measures the order parameter, which for a proper ferroelectric is the spontaneous polarization P_s. The *related* optical quantity is usually the birefringence, which in this case is proportional in lowest order to P_s^2. Moreover, the soft mode is Raman active only below T_c. Thus, the optical studies possible in these materials are not as complete as in the CJT ferroelastic. Nevertheless, much important information is to be gained. In fact, application of an electric field has been employed to lift the selection rules and allow observation of the soft IR active mode by Raman scattering above T_c in $SrTiO_3$ and $KTaO_3$ [2.66].

A number of familiar materials fall into this class, including $BaTiO_3$, $LiNbO_3$, and SnTe, to mention a few. The example we have chosen for this class of phase transition is $Pb_5Ge_3O_{11}$. It is a well-studied material, having been investigated by a variety of techniques (see [2.67] for a review of these). It undergoes a displacive proper ferroelectric transition at 451 K from the space group C_{3h} to C_3. (The C_{3h} assignment in the prototypic phase remains somewhat in doubt due to anomalies in the birefringence [2.68] and other properties [2.69] above T_c). The structural change at the phase transition is accompanied by a rotation of the GeO_4 tetrahedra about the ferroelectric axis. This imparts a "handedness" to the structure which is reflected in optical activity, reversible by polarization switching. This handedness gives rise to an additional optical method of investigating the static order parameter, the optical activity. Investigation of the optical activity, shown in Fig.2.9, as well as of the birefringence [2.70] have shown this transition to be second order.

The Raman scattering spectrum has been measured completely at room temperature and most of the lines observed have been assigned symmetry classifications [2.71].

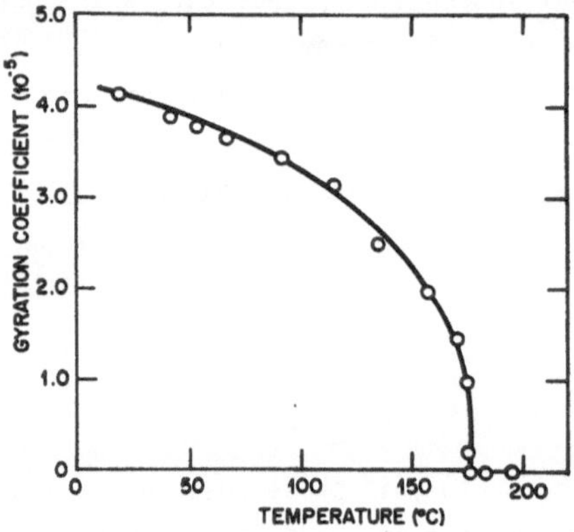

Fig. 2.9. The optical activity of $Pb_5Ge_3O_{11}$ near the phase transition at 451 K [2.70]

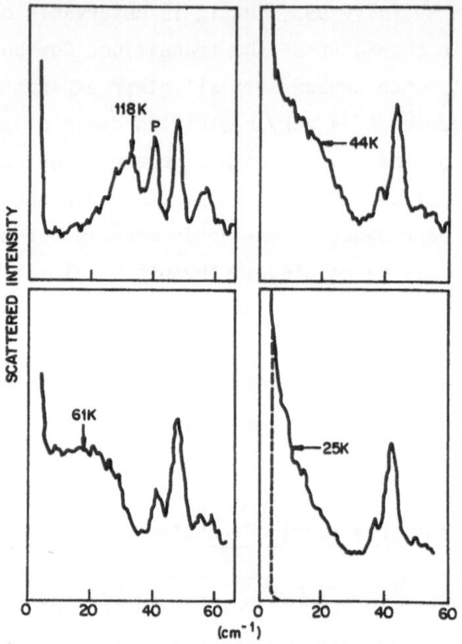

Fig. 2.10. Raman spectra of lead germanate showing the soft mode evolution (arrows) well below $T_c = 451$ K. The interaction between the soft mode and the 40 cm^{-1} phonon labeled "P" is not important for the dynamics close to T_c. The temperatures shown in the figure indicate the values of $(T_c - T)$

The soft mode observed in $x[zz]_y$ geometry below T_c has been followed over a wide range of temperature by various techniques. The first stage of its behavior occurs [2.72,73] for temperatures well below $T_c[(T_c - T) > 100$ K]. Its behavior in this region is complicated by interaction with two nontemperature-dependent optic modes. However, analysis of the spectra clearly shows that the soft mode obeys MFT in this region. Its characteristic frequency behaves as $\omega_s^2 \propto (T_c - T)$. Its intensity is relatively constant. It becomes overdamped at approximately $T_c - T = 40$ K, and its linewidth subsequently evolves as $\Gamma_{SM} \propto (T_c - T)$. Typical Raman spectra in this region are shown in Fig.2.10.

Within 1.5% (6 K) of T_c, this simple behavior starts to change. Study of the soft mode in this region was made possible [2.74] by the iodine cell technique mentioned in Sect.2.2.3. Two effects cause departures from the simple soft mode behavior observed well below T_c [2.67]. First, the mode interacts with acoustic modes as its frequency lowers. This interaction is little different in principle from that with optic phonons at lower temperature. However, since the coupling is via the piezoelectric effect, independent measurements of the latter allow this interaction to be quantitatively calculated with no adjustable parameters. The expected spectrum may be calculated by the methods described in Sect.2.2.3 for coupled modes, using the quasi-harmonic approximation for both the acoustic and soft optic modes. By such a calculation excellent agreement is obtained with the temperature dependence of the acoustic phonon frequency and linewidth, both of which exhibit small anomalies near T_c. However, the lineshape of the soft mode is not explained by this interaction. It can be explained, however, by lifting the quasi-harmonic assumption for the soft mode. This is effectively the same as assuming a coupling of the soft mode to a relaxation mode. The resulting susceptibility for the soft mode is [2.67]

$$\chi_s^{-1} = \mathrm{const}[\omega_{\underline{q}}^2 - \omega^2 - 2i\Gamma_{\underline{q}}\omega - i\Sigma(\underline{q},\omega)] \quad . \tag{2.30}$$

The self-energy $\Sigma(\underline{q},\omega)$ is given by

$$\Sigma = \frac{\omega\delta^2\tau}{1-i\omega\tau} \quad , \tag{2.31}$$

where τ is the relaxation time of the mode and δ is the strength of the coupling to the soft mode. This modified susceptibility is then used in (2.21) and (2.19) to calculate the spectrum. The result is the complicated lineshape shown schematically in Fig.2.11. The Brillouin acoustic peaks are designated B. The soft mode shows *two* characteristic frequencies now, one related to the high-frequency wing (W) and one to the dynamic relaxation peak (D). The detailed temperature evolution of the features D and B is shown in Fig.2.12, where the lines represent a calculation which includes both effects mentioned above. The experimental spectra, represented by the points, are well described by the model, where the (adjustable) relaxation frequency τ^{-1} of 29 GHz was found to give the best fit to the data. This relaxation frequency is similar to that expected for phonon density fluctuations [2.75] and hence the central component is attributed to that process. Although τ^{-1} is constant, the apparent central peak width narrows as $T \rightarrow T_c$, as can be shown mathematically from (2.30,31) above. This represents the first time that the "central peak" feature associated with a phase transition in a crystalline solid has been resolved and shown to exhibit a finite, temperature-dependent linewidth.

This already complex situation in $Pb_5Ge_3O_{11}$ is further complicated by the presence of an anomalous *elastic* scattering (zero linewidth) near T_c. This scattering must be of static origin, and thus is designated S in the figure. Without the use of a resonant reabsorption technique (Sect.2.2.3) it may completely prevent [2.76] the observation of the dynamic component. Through a combination of high-resolution Fabry-Perot interferometry [2.67] and autocorrelation spectroscopy [2.76] the elastic component has been shown to have a linewidth less than 10 Hz. The elastic intensity

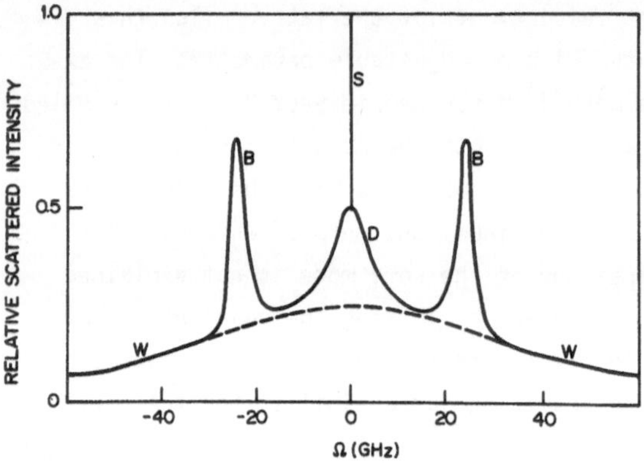

Fig. 2.11. Schematic of the coupled mode spectrum of $Pb_5Ge_3O_{11}$, showing the various distinguishable spectral features (see text)

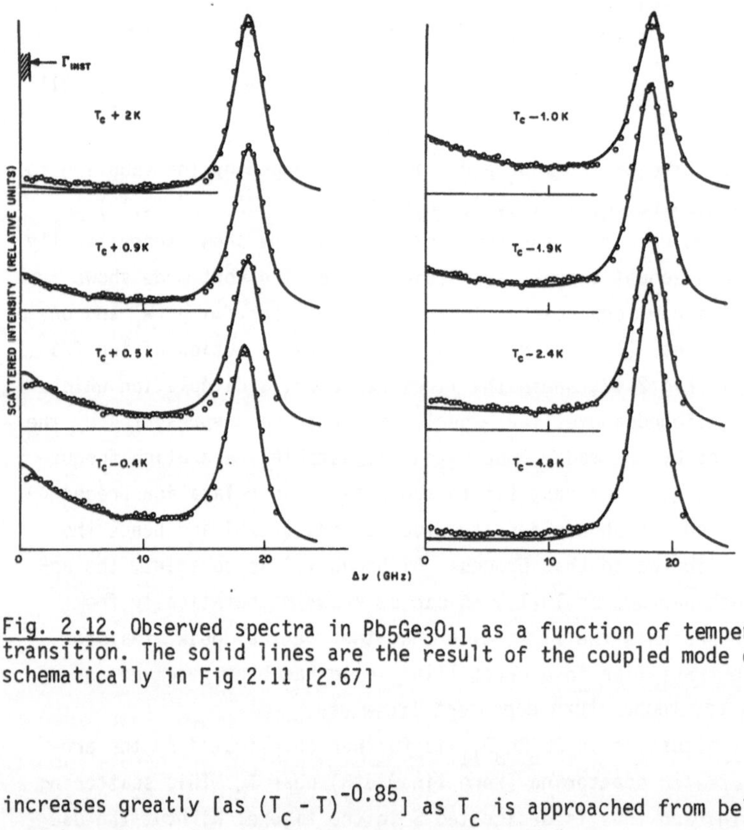

Fig. 2.12. Observed spectra in $Pb_5Ge_3O_{11}$ as a function of temperature near the phase transition. The solid lines are the result of the coupled mode calculation shown schematically in Fig.2.11 [2.67]

increases greatly [as $(T_c - T)^{-0.85}$] as T_c is approached from below and then rapidly vanishes above. The dynamic portion of the spectrum in the other hand, behaves as expected for a uniaxial ferroelectric - namely, its intensity undergoes a weak apparently logarithmic divergence, in violation of MFT (see Sect.2.2.3) but in qualitative agreement with the expectations of renormalization group theory [2.20]. The behavior of the intensity of the elastic peak seems to strongly suggest scattering from static defects. However, a simple defect model for such a dipolar system fails to predict such a strongly divergent intensity [2.67]. Whether the presence of these defects in turn modifies the observed dynamic properties of the soft mode is as yet an unanswered question.

In this case then, optical techniques, particularly light scattering, have been used to obtain a complete picture of the evolution of the soft mode dynamics as a function of temperature. This evolution is more complex than would have been anticipated and poses theoretical questions as yet unanswered.

2.5 Piezoelectric-Ferroelectric Transition in KDP and Its Isomorphs

Although the structural phase transitions in the AB_2CO_4 type materials (where A = K, Rb, Cs; B = H or D; and C = P or As) are first order, this family has received much attention in the literature. From a theoretical point of view the phase transition, from the piezoelectric but paraelectric D_{2d}^{12} structure to the ferroelectric C_{2v}' structure, offers at first sight an attractive model system, owing to the order-disorder character of the proton (or deuteron) motion involved. The displacements accompanying the transition, shown in Fig.2.13 [2.77], involve cooperative motions which can be viewed as proton tunneling through the mid-hump of a double well potential. The fact that this motion is perpendicular to the direction of the spontaneous polarization (and the displacement of the P ion) already suggests that the uncoupled tunneling mode does not describe the dynamics of this transition. Additional evidence for the importance of mode coupling arose from the discrepancy between the

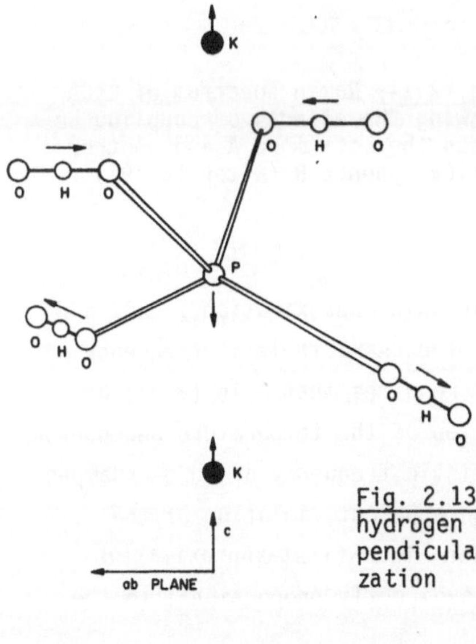

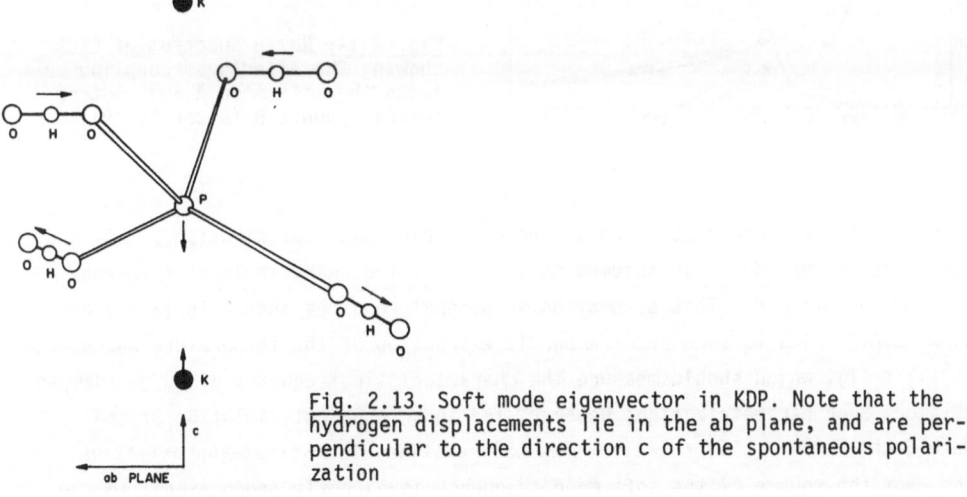

Fig. 2.13. Soft mode eigenvector in KDP. Note that the hydrogen displacements lie in the ab plane, and are perpendicular to the direction c of the spontaneous polarization

temperature-dependent soft mode frequency $\omega_0 = A[(T-T_0)/T]^{\frac{1}{2}}$ inferred from the original Raman data [2.78] and the predictions of soft mode theory. The complexities of the coupled mode situation are indicated by the widely differing interpretations which have been drawn from the Raman spectra of KDP and its isomorphs [2.78-82].

Despite these numerical differences, essentially all authors agree that a) above T_c the soft (tunneling) mode is linearly coupled to a slightly higher frequency optic phonon of the same (B_2) symmetry; b) because the crystals are piezoelectric above (as well as below) T_c, a piezoelectric coupling of the hybridized soft mode with the transverse acoustic (c_{66}) phonon occurs; and c) in some family members additional "central peaks" are seen very near T_c. Let us consider these phenomena in turn.

2.5.1 Coupled Modes in the Raman Spectrum

The importance of "proton-phonon" coupling was first emphasized by KATIYAR et al. [2.79] in the cases of CsH_2AsO_4 (CsDA) and KH_2AsO_4 (KDA). As shown in Fig.2.14 the observed spectral lineshape can be quantitatively represented by considering two coupled modes (whose shapes in the absence of coupling appear as dashed curves).

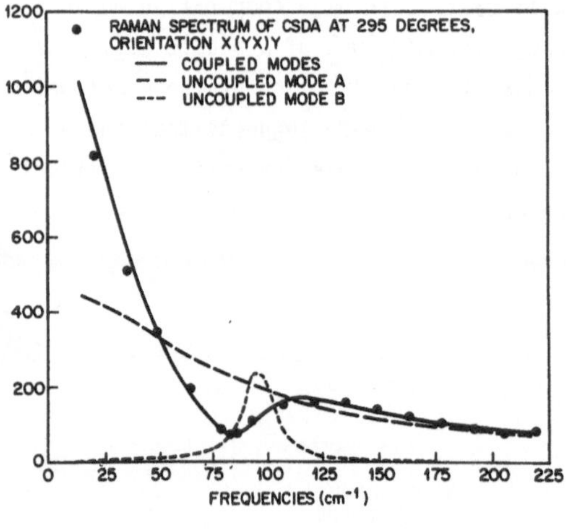

Fig. 2.14. Raman spectrum of CsDA showing the effects of coupling between the soft mode A and another optical phonon B (after [2.79])

Both uncoupled modes are considered in the quasi-harmonic approximation [2.20] and the temperature dependence is assumed to lie in ω_s, the characteristic frequency of the uncoupled soft mode. This assumption in general requires that A in (2.22) be complex [2.81]. Fits to such spectra permit extraction of the temperature dependence for $\omega_s^2(T)/\Gamma_s(T)$, which should measure the characteristic frequency of an overdamped oscillator. Such parametrization, however, led to an apparent violation of the Lyddane-Sachs-Teller (LST) relation. The LST relation, to a first approximation, states that the square of the soft mode frequency is directly proportional to the inverse dielectric susceptibility. Since the vanishing of the latter determines the ferroelectric transition temperature (T_c), the temperature at which the soft mode frequency vanishes (T_0) should be identical to T_c. The above coupled mode analysis often produced quite substantial values for this difference, $\alpha \equiv (T_c - T_0)/T_c$. For example in CsDA, α as large as 0.5 has been obtained [2.79], far in excess of that attributable to the quite modest first-order character (≈ 1 K) of the transition. COWLEY and COOMBS [2.75], realizing that an experimental observation which violates LST simply means that the full zero frequency response of the system has not been measured, proposed the modified form of χ_s in (2.30,31). The parameter δ^2 in (2.31), which may be temperature dependent, is related to α above. That is, if the soft mode frequency behaves as $K(T - T_0)$ we then find $\alpha = \delta^2/KT_c$. The values of α inferred by this analysis depend, of course, on the admittedly imprecise determination of ω_s. Hence, for CsDA values of α in the range 0.2 to 0.5 were found [2.83].

2.5.2 Brillouin-Raman Spectra

An intriguing consequence of the coupling discussed above is the existence of an additional *central peak* in the Raman spectrum whose strength scales as $\delta^2/[\omega_0^2(\omega_0^2 + \delta^2)]$, where $\omega_0^2 = \omega_s^2 - \delta^2$, and whose width should be roughly $\tau^{-1}\omega_0^2/(\omega_0^2 + \delta^2)$. Even the smallest value of α quoted above implies that significant intensity should lie in such a peak. However, a careful study under high resolution by LAGAKOS and CUMMINS [2.81] in CsDA revealed *no* such central peak, and permitted an upper limit on α to be set at $\lesssim 0.02$. The important extra ingredient which they introduced was the strong interaction between the soft mode and the transverse acoustic phonon (described by the c_{66} elastic constant), resulting in a *three*-coupled mode analysis. This interaction had been previously studied within a two-coupled mode formalism [2.84-86]. In this earlier work it was found to be responsible for a striking anomaly in the shear elastic constant near T_c, shown in Fig.2.15. LAGAKOS and CUMMINS [2.81] showed that, within the framework of the three-coupled mode analysis, the difference $(T_c - T_0)$ was essentially equal to the difference between the clamped (T_c^x) and free (T_c^σ) Curie temperatures,

$$(T_c^\sigma - T_c^x) = (a_{36}^T)^2 \frac{C}{4\pi} \frac{1}{c_{66}^{P,T}} \quad , \qquad (2.32)$$

where a_{36}^T is the piezoelectric coefficient, C the Curie constant, and $c_{66}^{P,T}$ the shear elastic constant under constant polarization. Thus, the apparent LST violation implied by the earlier two-coupled mode analysis of Raman spectra is more properly explained by the fact that the soft mode Raman frequency is sensitive to the *clamped* response, whereas the T_c observed in the experiments is determined by the *free* (zero stress) response. The invocation of a relaxing self-energy as in (2.30) for CsDA is therefore unnecessary.

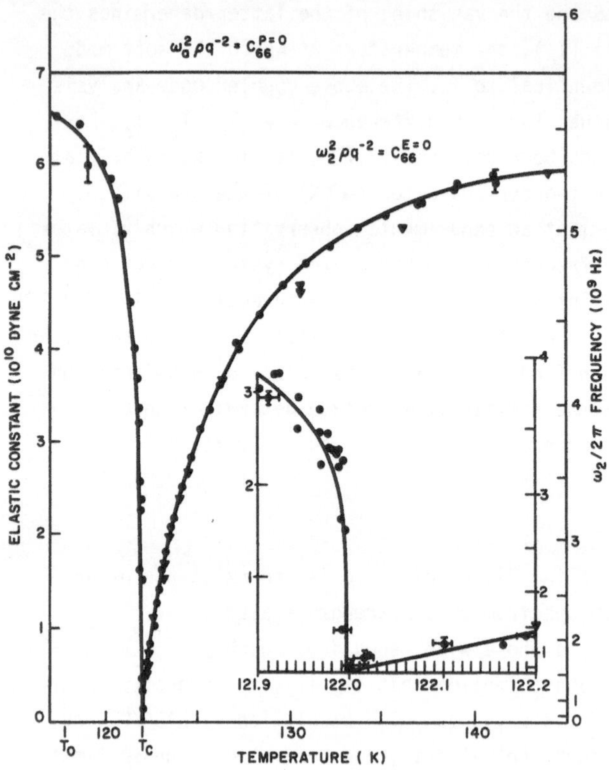

Fig. 2.15. Anomalous elastic behavior of c_{66} near the ferroelectric transition in KDP, observed by Brillouin scattering [2.84]

2.5.3 Static and Dynamic Central Peaks

It should be noted here that such relaxing self-energy components and their associated central peaks have quite recently been verified in the cases of $Pb_5Ge_3O_{11}$, $SrTiO_3$, and KDP, although the microscopic mechanisms responsible are not thought to be the same in all three cases. As discussed in Sect.2.3, the so-called "central peak problem" refers to the spectral intensity centered at zero frequency shift observed (in both neutron and light scattering experiments) in the vicinity of a structural phase transition, and which is in addition to the inelastic or quasi-elastic scattering attributable to the quasi-harmonic soft mode. The first reported instance of this effect for light scattering was made for quartz by IAKOVLEV et al. [2.87] and for neutron scattering by RISTE et al. in $SrTiO_3$ [2.88]. The quartz phenomenon was later shown [2.89] to be due to scattering from static microdomains, rather than to true critical opalescence. In most other cases, confusion still persists because of the lack of sufficiently sensitive spectral resolution to decide experimentally whether the central peak is a static or "dirt" phenomenon, or a dynamic process.

Observed divergences in scattered intensity near T_c are not sufficient in these cases. A definitive determination of the *spectral* content of the central peak is required. Such a diverging central peak has been observed in KDP above T_c [2.81] but was shown by speckle photography [2.90] to be largely of static origin. The uniform, albeit static, spatial appearance of the speckle suggests that finely distributed immobile defects are present. COURTENS [2.91] has explored the possibility that deuterium in its naturally occurring abundance of 0.02% could be responsible. His latest results [2.91a], however, indicate that the central peak disappears, after annealing, suggesting that the defects responsible for it are not isotopic impurities.

Of greater interest from the dynamical point of view are the very recent high-resolution experiments in KDP below T_c by MERMELSTEIN and CUMMINS [2.92] which have revealed a diverging central peak of *finite* linewidth. Seen only within 0.1 K of the transition, this feature has a width of ≈ 50 MHz, independent of temperature within experimental accuracy, and apparently arises from the coupling between the soft ferroelectric mode and the thermal diffusion mode.

It is generally true that in the ordered phase a linear coupling between the temperature T and the order parameter ψ will always exist. In particular, the scattering of light by entropy or temperature fluctuations is proportional to

$$\langle \varepsilon_{\alpha\beta}^2 \rangle = \frac{1}{V}\left[\frac{\partial \varepsilon_{\alpha\beta}}{\partial T}\right]_P^2 \frac{kT^2}{\rho C^P} \tag{2.33}$$

and

$$\left[\frac{\partial \varepsilon}{\partial T}\right]_P = \left[\frac{\partial \varepsilon}{\partial T}\right]_{X,P} + \left[\frac{\partial \varepsilon}{\partial X}\right]_{T,P} \frac{\partial X}{\partial T} + \left[\frac{\partial \varepsilon}{\partial P}\right]_{T,X} \frac{\partial P}{\partial T} \quad , \tag{2.34}$$

where X and P label strain and polarization variables, respectively. The first term in (2.34) describes the direct scattering recently observed by LYONS and FLEURY [2.93] and is quite small for most solids. The second and third terms describe indirect coupling of light to temperature fluctuations via temperature dependence of spontaneous strain and polarization below T_c. In the absence of dynamic coupling effects the lineshape of the entropy fluctuation scattering is a Lorentzian of width $\Gamma = q^2(\Lambda/\rho C^P)$, centered at zero frequency shift. Here $\underline{q}$ is the scattering wave vector and $\Lambda/\rho C^P$ the thermal diffusivity. The known thermodynamic parameters for KDP predict a linewidth of 62 MHz for right-angle scattering. The observed 47 ± 5 MHz value obtained experimentally argues strongly for this identification of the dynamic portion of the central peak [2.92]. The temperature dependence of the scattered intensity is determined by the temperature derivatives of X and P_s. For KDP both of these vary as $(T_c - T + 0.026\ \mathrm{K})^{-5/6}$ and are strongly divergent. The strength of this divergence, together with the fact that the phase transition is very slightly first order, precluded a quantitative comparison of experimental intensities with the theory. Also, due to experimental constraints [2.92] it was not possible to verify the q dependence of the central peak linewidth. Nor was the critical narrowing of the central peak, according to $\Gamma = \Gamma_{th}[\omega_0^2/(\omega_0^2+\delta^2)]$ as predicted for this coupling, observed. Since the ferroelectric transition in KDP has been found to become second order at $\approx 2.0 \pm 0.3$ kbar [2.94,95], experiments at elevated pressures, as well as at other scattering angles, might permit quantitative verification that this central peak is due to entropy fluctuations.

2.6 Zone Boundary Phase Transitions (Antiferrodistortive)

The importance of soft modes occurring for wave vectors other than $\underline{q} = 0$ is by now widely appreciated. The physical significance of nonzero $\underline{q}_\psi$, of course, is that the atomic displacement amplitude is modulated from unit cell to unit cell with a spatial periodicity equal to $2\pi/\underline{q}_\psi$ in the direction $\hat{\underline{q}}_\psi$. We now know that $\underline{q}_\psi$ need not even be commensurate with the periodicity of the prototypic lattice. We shall discuss these interesting "incommensurate" transitions in Sect.2.8. In the present section we shall discuss the commensurate unit cell multiplying transitions as exemplified by the perovskite family.

Perhaps the best studied perovskite is $SrTiO_3$, which undergoes a cubic-tetragonal phase transition at 106 K, associated with the condensation of a triply degenerate R point phonon. As shown in Fig.2.5, the distortion from the cubic ABO_3 perovskite structure can be regarded as antiphased rotations of the BO_6 octahedra in adjacent cubic unit cells. Ignoring coupling to strain [2.96] or to electronic levels [2.7], the free energy associated with this distortion may be written [2.97] in terms of ψ_x, ψ_y, ψ_z, the amplitudes of the octahedral rotational components about the x, y, or z axes,

$$V = \frac{a}{2}(\psi_x^2+\psi_y^2+\psi_z^2) + \frac{b}{4}(\psi_x^4+\psi_y^4+\psi_z^4) + \frac{c}{2}(\psi_x^2\psi_y^2+\psi_x^2\psi_z^2+\psi_y^2\psi_z^2) \quad . \qquad (2.35)$$

This form, applicable to all perovskites, may be minimized for three types of distortion, tetragonal, orthorhombic, and trigonal, corresponding, respectively, to $\psi_x \neq 0$, $\psi_x = \psi_y \neq 0$, and $\psi_x = \psi_y = \psi_z \neq 0$. The nonzero components of ψ, below T_c play the role of the microscopic order parameter. The stable phase is determined by the relationship among the coefficients a, b, and c [2.97]. Thus from (2.35) it has been shown that the orthorhombic distortion is never the stable phase. In practice this phase can be stabilized by couplings not included in (2.35), such as the CJT interaction in $PrAlO_3$, discussed in Sect.2.3. The tetragonal phase is stable when $c > b$, while the trigonal distortion is favored for $-\frac{1}{2} < c/b < 1$. This simple model

[2.35] accounts at least qualitatively for a remarkable amount of data on the perovskite cell doubling transitions.

2.6.1 Systematics in the Fluoroperovskites

Before discussing $SrTiO_3$ in particular, we shall now mention some of the wide variety of work on other perovskites. ROUSSEAU and co-workers [2.98] have performed a coordinated series of crystallographic, ultrasonic, and light scattering studies for several members of the fluoroperovskite family, including the AMF_3 series (where A = K, Rb; M = Mg, Ni, Co, Zn, and Mn) as well as the $BCdF_3$ series (where B = Rb, Cs, or Tl). This highly ionic series of crystals provides an opportunity to understand why some fluoroperovskites exhibit the cell doubling phase transition while others do not, and avoids the difficulties of the anomalous oxygen polarization. From an analysis of a simple force constant model and a consideration of ionic radii and lattice parameters, ROUSSEAU et al. [2.98] expected a structural transition due to rotations of the MF_6 octahedra to occur in those fluoroperovskites where the cubic lattice parameter a is greater than $(\underline{r}_A + \underline{r}_F)\sqrt{2}$. Here $\underline{r}_A$ and $\underline{r}_F$ are the radii of the A and F ions. This simple rule predicts correctly the fact that the phase transition occurs in relatively few fluoroperovskites, as well as its nonoccurrence in many others.

Raman [2.99] and Brillouin spectra for $KMnF_3$ near its first-order transition at 186 K have been found to behave similarly to those of $SrTiO_3$. Birefringence [2.100] in $KMnF_3$ has revealed no fluctuating component above T_c (such as was found by COURTENS in $SrTiO_3$) but is consistent with a value of $\beta \approx 0.33$ for the order parameter critical exponent below T_c.

Both $RbCdF_3$ (T_c = 124 K) and $TlCdF_3$ (T_c = 191 K) exhibit the $O_h^1 \to D_{4h}^{18}$ $SrTiO_3$ transition, and both the A_{1g} and E_g soft modes have been studied by Raman scattering [2.98]. Strong damping of the soft modes near T_c has precluded any attempt to extract critical exponents or to shed light on the possible existence of a central peak for these transitions. Striking anomalies in the ultrasonic velocities on the other hand suggest that fluctuation effects may be significant above T_c. In both materials the elastic constants exhibit appreciable softening (several percent) well above T_c. In $RbCdF_3$ c_{11} shows evidence of the steplike behavior familiar in $SrTiO_3$ (see Sect.2.6.2) superimposed upon the sharp velocity minimum. So far data on other elastic constants have not been obtained below T_c. Quite recently high-resolution Brillouin spectra [2.101] of $TlCdF_3$ and $RbCaF_3$ have revealed both attenuation and elastic constant anomalies strongly suggestive of order parameter fluctuation effects as far as 80 K away from T_c. In particular, a large velocity dip of the type predicted by PYTTE [2.102] has been observed. These results emphasize the need for central peak studies in these materials of the type reported for $SrTiO_3$ [2.103].

Other haloperovskites examined by optical techniques include $TlMnCl_3$ [2.104] and $RbCaF_3$ [2.105]. The former apparently traverses the structural sequence: cubic → tetragonal → orthorhombic → monoclinic at 296 K, 276 K, and 235 K, respectively. In $RbCaF_3$ the Raman studies [2.106] suggest that the transition at 193 K is of the same type as in $SrTiO_3$.

2.6.2 Recent Results in $SrTiO_3$

Although $SrTiO_3$ was the first verified example of a zone boundary soft mode transition, it has provided the most fertile ground for demonstrating the shortcomings of the simple soft mode theory. There have been two, not necessarily related, types of shortcoming manifested in various experiments. The first can be described as a failure of mean field theory to predict correctly the critical behavior of the order parameter and its fluctuations. Beginning in 1971 MÜLLER and BERLINGER [2.107] have thoroughly demonstrated that close to T_c the order parameter as probed by EPR spectra from various impurity centers fails to evolve as $t^{1/2}$ but rather exhibits a β close to 0.33. The many subtleties of these experiments including possible crossover effects and multicritical phenomena associated with strains will be treated by Müller's chapter in Vol.2 of this series. The basic behavior is observed in the bulk crystal via birefringence measurements [2.6]. The failure of mean field theory for a structural transition is not unexpected, and does not impune even the quasi-harmonic soft mode concept. Naturally, if mean field theory fails and $\beta \neq 1/2$, one also expects that the exponent describing the soft mode temperature dependence would depart from its predicted mean field behavior, as discussed in Sect.2.2.3.

Since T_c in $SrTiO_3$ is known to be sample dependent (ranging from 90 to 110 K), any exponent extraction must include a precise determination of T_c in the sample under study. Even if T_c is known precisely, the characteristic frequency of the order parameter fluctuations can be easily extracted from the light scattering spectrum only in the simplest case. In the presence of mode coupling or anharmonic effects, the entire spectrum, including any central peak, must be considered in the extraction of ω_ψ and its associated critical exponent z, as shown by (2.25). It is possible that, over a given temperature region, observation of the high-frequency scattering alone can lead to the conclusion that $\upsilon z = \beta$, but this equality cannot be regarded as a true measure of the dynamic critical exponent. The observation by STEIGMEIR et al. [2.108] that $\omega_s^2 \alpha t^{0.66} = t^{2\beta}$ must be viewed in this light, even if the uncertainty in T_c is ignored. We note, however, that an earlier Raman investigation of the same temperature region [2.109] demonstrated that the apparent soft mode exponent could be varied between 0.5 and 0.33 simply by adjusting T_c by less than 3 K.

The second shortcoming of the simple soft mode picture discovered in $SrTiO_3$ lies in the so-called "central peak problem". The general problem was stated in Sect.2.4. In the intervening years since the initial report in $SrTiO_3$ by RISTE et al. [2.88], virtually every structural phase transition examined by neutron scattering has exhibited a central peak. Mechanisms considered for this phenomenon include selective phonon interactions, dynamic clusters, domain walls, static and mobile defects, and solitons [2.110-114]. EPR, X-ray, Mössbauer, neutron, and light scattering experiments have all probed for the central peak. The only scattering experiment to measure a finite linewidth (i.e., a dynamic component) to the central peak [2.103] has also inferred the existence of a much narrower and possibly static component. The most recent neutron study [2.115] has deliberately varied the concentration of defects and has demonstrated a (highly sublinear) correlation with observed central peak intensity. Furthermore, in both lead germanate and KDP, both static and static and dynamic components to the central peaks have been identified. These experiments taken together fortify the notion that there is no single type of central peak. Rather several mechanisms are sometimes even simultaneously at work. The static components appear to arise from defects and may be considered "extrinsic" to the critical dynamics. We also note that these defects may modify the dynamic order parameter fluctuation. The dynamic components appear to arise from anharmonic interactions of the soft mode with temperature or phonon density fluctuations and are "intrinsic" in the sense that they will always be present even in the experimentally inaccessible case of a perfectly pure and defect free solid.

Using a refined experimental technique described in Sect.2.2.3, LYONS and FLEURY [2.93] spectrally resolved both entropy fluctuation and phonon density fluctuation central peak components in the polarized spectra of $KTaO_3$ and $SrTiO_3$. They detected however, no appreciable singular behavior in either component near the cubic-tetragonal transition in $SrTiO_3$, despite the well-known collapsing A_{1g} soft mode component. Following this report, COURTENS suggested [2.116] that the depolarized (E_g symmetry) spectrum be examined more carefully in view of his earlier detection of fluctuation-dominated birefringence above T_c in $SrTiO_3$. COURTENS emphasized the connection between the fluctuating birefringence and the total depolarized scattering intensity due to *soft mode* phonon density fluctuations. The subsequent light scattering experimental results [2.103] did indeed reveal a singular dynamic central peak of E_g symmetry in $SrTiO_3$, probably associated with phonon density fluctuations. However, the situation is complicated by interactions with the transverse acoustic phonons. Careful lineshape analyses of the central peak-Brillouin spectra over the range $-15\ K < T - T_c < 5\ K$ showed that the observed central peak reached a maximum intensity approximately 1.5 K below T_c, and was not observed to narrow appreciably below ≈ 10 GHZ. These facts, together with the observed anomalies in the TA acoustic velocity and absorption (see Fig.2.16a), were all accounted for in terms of a coupled mode analysis quite similar to that discussed earlier for lead germanate (Sect.2.3). The uncoupled susceptibilities are as given in (2.20) (for the acoustic mode) and (2.30,31) for the soft mode. The scattered lineshape is then given by (2.21,22) where $A = a\psi$ determines the coupling parameter A in terms of ψ, the order parameter

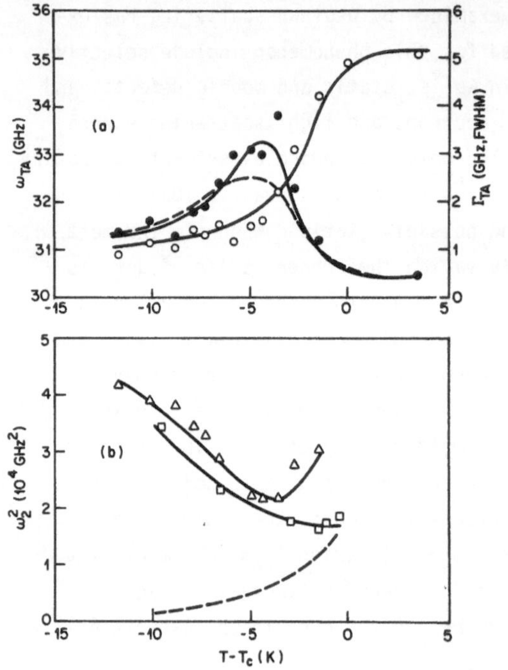

Fig. 2.16. (a) Brillouin scattering observations of the TA velocity (ooo) and attenuation (●●●) anomalies near T_c = 106 K in $SrTiO_3$, compared with a coupled mode calculation (---) for the attenuation. Solid lines are guides to the eye. (b) Temperature dependence of the soft mode relaxation parameters for $SrTiO_3$: δ^2 (---), ω_0^2 (□), and $\Delta = \omega_\infty^2$ (Δ) [2.103]

obtained from EPR data. The proportionality constant is a = 3100 GHz^2/K, chosen to agree with Brillouin data well below T_c. The fits to the spectral profiles produced $\tau^{-1} \approx 15$ GHz (temperature independent) and the behavior shown in Fig.2.16b for other quantities. Note in Fig.2.16a that both the velocity and absorption anomalies are correctly described. The overall lineshape, including the central peak, is also correctly given. From the resulting behavior of $\omega_{s0}^2 = \omega_s^2 - \delta^2$ and δ^2 it was concluded that: a) δ^2 reaches the same value upon approach to T_c from below as the neutron scattering found from above and b) the value for ω_{s0}^2 shown by squares in Fig. 2.16b does *not* reach zero at T_c, but saturates at $\omega_{s0}^2 \approx 1.5 \times 10^4$ $[GHz]^2$. This behavior implies the existence of an inaccessibly slow relaxation process which should give rise to an even narrower central component of comparable intensity. The iodine cell technique permits setting only an upper limit of ≈0.3 GHz on the linewidth of the second central component. The existence of two components in the $SrTiO_3$ central peak accounts at least qualitatively for all the experimental results reported thus far on the phase transition.

2.7 Zone Boundary Ferroelectrics: The Molybdates

In 1968 CROSS et al. [2.117] reported the observation of anomalous elastic and dielectric behavior at the ferroelectric-ferroelastic phase transition in $Gd_2(MbO_4)_3$. Although GMO was clearly ferroelectric below T_0 = 159 C, it exhibited no appreciable increase in the dielectric constant near T_0. Certain elastic constants, however, exhibited very large ($\geq$30%) anomalies. Early Raman [2.118] and IR [2.119] experiments showed no appreciable softening in any of the zone center phonon modes below T_0, although extremely temperature-dependent broadening of the lowest A_1 mode shifted the peak in the response severely toward zero frequency. Subsequent suggestions by PYTTE [2.120] and by LEVANYUK and SANNIKOV [2.121] that the GMO transition was associated with a zone boundary soft phonon were confirmed by the neutron studies in isomorphous $Tb_2(MoO_4)_3$ of DORNER et al. [2.122]. The elastic and ferroelectric behavior below T_0 was attributed to anharmonic interactions of the strain and polarization with the microscopic order parameter (MoO_4 rotations). In particular, the ferroelectricity results from a piezoelectric coupling to the strain produced by almost antiferroelectric tetrahedral rotations. DORNER et al. [2.122] followed the frequency decrease of the soft M point phonon from 750° C down to ≈375° C. Over this range the mode frequency is given by $\omega_m^2 = \omega_{m0}^2(T - T_c)/(T_0 - T_c)$ where ω_{m0}^2 = 0.165 meV^2. T_0 = 159° C, and T_c = 149° C. Close to T_0 the soft mode evolution was inferred from the integrated inelastic intensity under the assumption that $I \approx \omega_m^{-2}$ and an unresolved central peak was also observed. Some neutron measurements below T_0 were also performed, which produced results consistent with the earlier Raman and IR experiments we now discuss.

FLEURY [2.118] had observed the lowest frequency A_1 feature in the room temperature Raman spectrum of GMO to exhibit anomalous behavior as the temperature was raised toward T_0. Although of the correct symmetry to be the soft mode below T_0, this feature, when fitted to the quasi-harmonic oscillator response $\chi_s^{-1} = [\omega_s^2(T) - \omega^2 + i2\omega\Gamma_s]$, appeared to exhibit no appreciable softening, but rather a dramatic broadening upon approach to T_0. This broadening was responsible for the shift toward zero of the peak Raman scattered intensity. That is ω_s = 46 cm^{-1} ≈ const while $\Gamma_s(T)$ increased markedly near T_0. In addition, the second component expected to be split off by the degeneracy removal accompanying the orthorhombic distortion was not evident. SHEPHERD [2.123] extended these measurements by exploring the angular dependence of the Raman spectrum. As expected, due to the small infrared oscillator strength of the 46 cm^{-1} mode, no polariton effects were observed on the mode frequency. However, a still unexplained $\underline{q}$ dependence to the Raman linewidth was found. The linewidth varies by a factor of two in the range $0.2 \cdot 10^5 < \underline{q} < 10^5$ cm^{-1}, at constant temperature, between room temperature and T_0. Uniaxial stress experiments [2.124] showed as well that ω_s was insensitive to stress, although Γ_s proved quite sensitive to it.

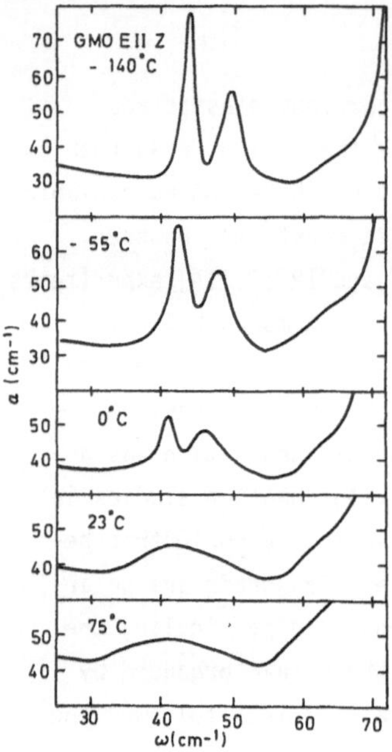

Fig. 2.17. IR absorption spectra in ferroelastic-ferroelectric GMO, showing the emergence at low temperatures of two A_1 components [2.119]

The missing second soft mode component below T_c was first identified in the IR experiments of PETZELT [2.119], who carried out measurements on GMO to well below room temperature. He found (Fig.2.17) that the A_1 mode studied by FLEURY split into two resolvable modes of unequal strenght below 0° C. Recently a detailed Raman study of these two modes was carried out by SHIGENARI et al. [2.125]. Above room temperature these data agree with the earlier results [2.118]. However, the lineshape is interpreted as a composite of two damped but noninteracting peaks rather than one. They fit their spectra to the form

$$I(\omega) = (n + 1) \left[\frac{K_1}{(\omega_1^2-\omega^2)^2+4\gamma_1^2\omega^2} + \frac{K_2}{(\omega_2^2-\omega^2)^2+4\gamma_2^2\omega^2}\right] \qquad (2.36)$$

extracting values for the parameters ω_1, γ_1, ω_2, and γ_2 for $T < 100°$ C. The parameter ω_2 does show appreciable softening, but the spectral lineshapes are such that the values of parameters emerging from this fit cannot be considered unique. This is emphasized by the equally good fit to a single damped oscillator previously achieved [2.118] between room temperature and T_0. Further, even at temperatures where ω_1 and ω_2 are resolvable, the scattering strength from the temperature-dependent mode 1 is an order of magnitude greater than from mode 2. Thus, the behavior of the two soft mode components below T_0 remains unresolved and the suggestion of a real mode softening unsubstantiated.

The mean field analysis of DORNER et al. [2.122] suggests that the minimum values of ω_1, ω_2 should be $\omega_{1,2} \leq 6$ cm^{-1}. This appears not to be the case. Another possibility which must be considered is that of a relaxing soft mode self-energy whose central peak could be responsible for the observed acoustic anomalies.

Several conventional ultrasonic and light scattering investigations of the anomalous *elastic* behavior in GMO have appeared [2.126-129]. The most dramatic behavior is seen in the longitudinal x-directed phonons. X refers here to the orthorhombic a axis. The corresponding elastic constant c_{11} determines the velocity $\rho v_{xx}^2 = c_{11}$ and, as observed in Brillouin scattering [2.130], is displayed in Fig.2.18. The behavior can be qualitatively described as the sum of a discontinuous decrease at T_0 and a singular "dip" as T_0 is approached from above as well as below. This behavior for very high-frequency (10-17 GHz) phonons is essentially identical to that reported by HÖCHLI [2.126] for 10-30 MHz phonons. Thus, any important relaxations appear to lie outside the 10 MHz-20 GHz range, judging from the sound velocity data. The information on sound absorption is less definitive. Both conventional [2.126,127] and Brillouin [2.127,130] studies show apparent attenuation peaks for α_{11} near T_0. However, the absolute values reported for the different frequency ranges do not yield a simple frequency dependence of the critical attenuation.

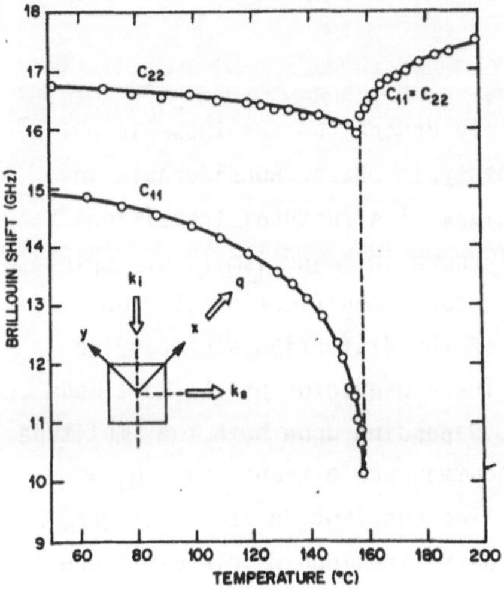

Fig. 2.18. Acoustic anomalies near $T_c = 159°$ C in GMO, as observed in Brillouin scattering. The elastic constants C_{ij} are proportional to the squares of the indicated Brillouin shifts [2.130]

COURDILLE et al. [2.128] have attempted to apply the theory of PYTTE [2.120] to describe their 540 MHz ultrasonic results in TMO. They conclude (in contrast to HÖCHLI) that the soft mode fluctuations are three dimensional in character and overdamped. However, the values inferred for the soft mode parameters ($\omega_s^2/\Gamma_s = 23.9$ GHz at 140° C) do not correspond to any features thus far observed spectroscopically. It should be noted that ITOH et al. have performed [2.127,130] high resolution double Fabry-Perot light scattering experiments on GMO and have not observed any such narrow soft mode or central peak component.

Thus, the picture in the improper ferroelectrics of the GMO type remains incomplete. Well above T_0 a quasiharmonic mean field like soft phonon has been observed at the tetragonal M point [2.122]. Below T_0, two low-frequency A_1 modes with anomalous temperature dependence have been observed [1.119,125]. Neither, however, softens sufficiently to account quantitatively for the observed acoustic behavior. Furthermore, the extreme broadening of these nearly degenerate modes precludes accurate separation of their contributions to the observed Raman and IR spectra. No dynamic central peak has been observed, but an unresolved central feature above T_0 was evident in the neutron spectra. Clearly, a high-resolution, high-contrast light scattering investigation of the type carried out for $SrTiO_3$ should be undertaken for these molybdates. A possible key to the behavior below T_0 may lie in a slight variation on the anharmonic couplings considered previously. Specifically, an extrapolation of the frequencies for the two A_1 components (44 cm^{-1} and 48 cm^{-1} at -100° C) to higher temperatures, where they can no longer be resolved, appears to indicate that $\omega_1 \to \omega_2$ for $T \approx T_0$. This leads one to speculate that the *frequency difference* between these two modes might play the role of the slow variable and that the nonlinear coupling of the two-phonon difference process to the strain might account for the observed acoustic behavior. More experiments are needed to decide.

2.8 Incommensurate Phase Transitions

Another class of phase transitions not yet completely understood are those in which the distorted phase is incommensurate with the prototypic phase. Considerable interest recently has arisen [2.17] regarding this class of structural transitions. Although examples of incommensurate order are well known in magnetism (e.g., spin density waves in chromium) their appearance as structural modifications is considerably less familiar. A schematic illustration of the distortion accompanying an incommensurate transition appears in Fig.2.19. The eigenvector of the soft mode δu corresponds to a two-component order parameter, depending upon both the amplitude and the phase of $\delta u(\underline{r}) = \delta u \exp(i\underline{q}_i \cdot \underline{r})$. For an incommensurate transition, $q_i = [2\pi(1-\delta)/na]$, where a is the prototypic phase lattice constant, n is an integer, and δ is an irrational number, small compared to unity. The long wavelength fluctuations about the new structure describing the excitations in the incommensurate phase can be decomposed [2.131] into that in which the *amplitude* $|\delta u|$ is uniformly changed and that in which the *phase* of δu is uniformly shifted $[\delta u \to \delta u \exp(i\phi)]$. The former or "amplitude" mode behaves as a normal soft mode in the ordered phase of a commensurate transition. The latter phase mode or "phason" behaves differently. Because the incommensurate distortion can be uniformly displaced by an arbitrary amount without cost in energy, the frequency of the q = 0 phason must be zero. This argument suggests that in the incommensurate phase, a new hydrodynamiclike excitation

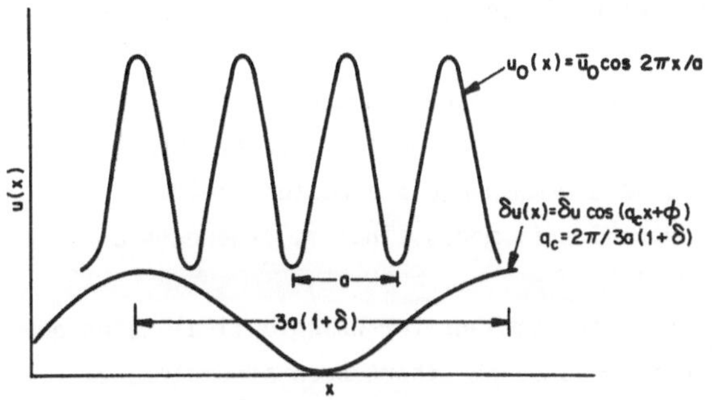

Fig. 2.19. Schematic representation of an incommensurate phase transition. The static particle displacements in the parent phase (u_0) are shown, together with the incommensurate distortion δu. The incommensurate structure is determined by the sum $u_0(\chi) + \delta u(\chi)$

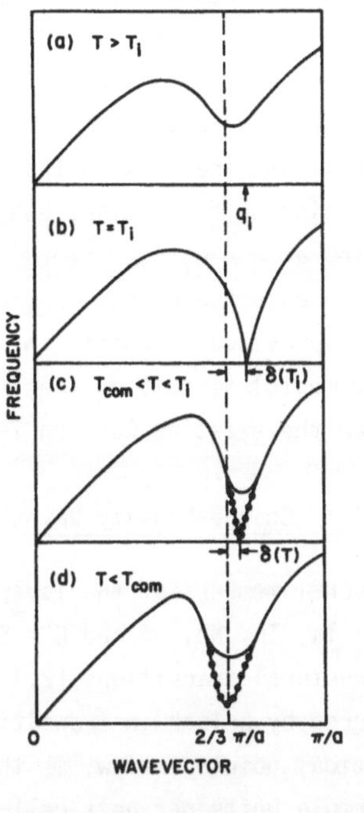

Fig. 2.20a-d. Schematic behavior of the soft mode dispersion curves above, at, and below the incommensurate transition (T_i) in an extended zone scheme, showing the amplitude (——) and phase (••••) components

exists - similar to a fourth acoustic mode. Within mean field theory one may argue further that the phason damping must vanish as $\underline{q} \to 0$. COWLEY and BRUCE [2.132] have argued that the incommensurate structural phase transition may be mapped into the x-y model, familiar in magnetic critical phenomena. The excitations in this model, the longitudinal and transverse spin fluctuations, behave somewhat differently than predicted by the simple schematic arguments above. To date, however, no direct observation of the "phason" in the incommensurate phase has been made. The mean field expected behavior is sketched in Fig.2.20.

Below $T = T_i$ the point $\underline{q}_i$ becomes a new reciprocal lattice point so that the new excitations appear at $\underline{q} = 0$ and may be probed by long wavelength techniques like optical scattering and absorption, ultrasonics, etc. We have not shown this effective folding back of the dispersion curves in Fig.2.20 to avoid confusion. However, it is clear that such multiple folding back within a repeated zone scheme would in principle cause many dispersion curve crossings and hence possible interactions which could alter the dynamics of the incommensurate phase excitations. Such interactions have not been calculated and experimental observations which bear on them have barely begun.

In some cases [2.133,134] evidence for an evolution below T_i toward a commensurate structure exists. That is, $\underline{q}_i(T) \to 2\pi/na$, or $\delta \to 0$. To date, all observed instances of this "lock-in" transition have been first order. However, it has been pointed out [2.132] that the lock-in transition need not be first order. Whether or not it is, the lock-in transition results in a *commensurate* structure below T_{com}, wherein the excitations behave again as normal soft modes. That is, there is a "gap" at $\underline{q} = \underline{q}_i$.

To date there are very few incommensurate transitions for which optical techniques have provided essential information. And in no case has there been a complete experimental assault on the incommensurate excitations. In this section we shall review the existing work on $TaSe_2$, K_2SeO_4, and $BaMnF_4$.

2.8.1 Charge-Density Waves: $TaSe_2$

Several members of the layered transition metal dichalcogenide family MC_2 (where M = Ta, Ti, Nb, Mo and C = S, Se) have been discovered to exhibit incommensurate structural transitions [2.131,135,136], some of which are followed at lower temperatures by a lock-in transition. The most thoroughly studied, certainly from the optical point of view, is the 2H polytype of $TaSe_2$. This polytype contains two $TaSe_2$ formula units per unit cell with the Se ions in a trigonal prismatic coordination.

At T_i = 122 K a second-order structural transition to an incommensurate state occurs, with the formation of a *nearly* 3a superlattice. The corresponding critical wave vector is $\underline{q}_i = (1-\delta)a^*/3$, with $a^* = 4\pi/a\sqrt{3}$ and $\delta(T_i) = 0.02$ and temperature dependent [2.134]. Although no simple soft mode behavior has been observed around $\underline{q}_i$ above T_i, quasi-elastic critical scattering of neutrons (i.e., a central peak) has been seen. Further, the value of $\underline{q}_i$ appears to be determined by $\underline{q}_i = 2\underline{q}_F$ where $\hbar\underline{q}_F$ is the electron Fermi momentum. Thus, the incommensurate transition in the MC_2 compounds is driven by the electron-phonon interaction so that the distortion below T_i can be represented by a static charge density wave (CDW). Most of the theory and experiments on CDWs do not concern optical processes, so we refer the reader to the literature [2.17] for these details. At least three groups have performed Raman scattering experiments [2.137-139] on 2H-$TaSe_2$ which have provided increased understanding of the CDW excitations. Above T_i the only temperature-dependent Raman feature is a broad band at $\approx 130\ cm^{-1}$ which softens very slightly and weakens, but persists down to at least 89 K (i.e., below T_{com} = 90 K). This has been identified as second-order scattering from the 7 meV, $k \approx \underline{q}_i$ phonons observed in neutron scattering. Below T_i several new low-frequency lines emerge. Upon still further cooling, the first order lock-in transition occurs at T_{com} = 90 K. Figure 2.21 shows the temperature evolution of the low-frequency Raman spectrum [2.138] in 2H-$TaSe_2$. The $49\ cm^{-1}$ E_{2g} mode softens upon approach to T_i from below and behaves as expected for the amplitude mode. The softening A_{1g} mode at $82\ cm^{-1}$ cannot be seen above T_{com} and is probably the soft mode which becomes the phason between T_{com} and T_i. This inter-

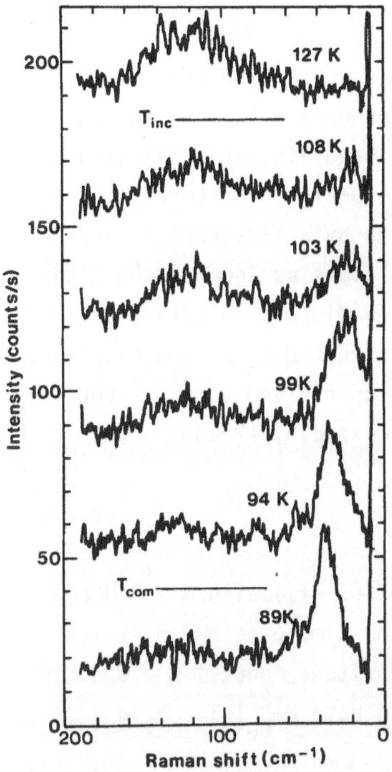

Fig. 2.21. Raman spectra of CDW related excitations in 2H-$TaSe_2$, showing the softening of the ~40 cm^{-1} mode as T is raised toward T_c. Note that frequency increases from right to left (after [2.138])

pretation however disagrees with that of HOLY et al. [2.137] who assigned the lowest A_{1g}(44 cm) and E_{2g} (49 cm^{-1}) modes as phase modes and the higher lying A_{1g} (82 cm^{-1}) and E_g (65 cm^{-1}) modes to amplitude modes. Additional arguments as well as data for IT-$TaSe_2$ appear in TSANG et al. [2.139]. The situation is complicated by possibilities of interactions among the CDWs which have not yet been fully accounted for even by the recent microscopic theory of McMILLAN [2.140]. Further work by McMILLAN [2.141], involving interaction of the CDWs with the host lattice and the concommitant tendency toward discommensurations, suggests the existence of a gap in the phason dispersion curve at the first "Brillouin zone boundary" of the incommensurate lattice and, further, demonstrates the possibility that such interactions may prevent the $\underline{k} = 0$ phason from continuously approaching zero as the lock-in transition is approached from below. These interesting suggestions are not yet verified experimentally. Further, the failure to observe the phason mode in the incommensurate state leaves undetermined some important parameters in the theory. Finally, low-frequency acoustic studies in 2H-$TaSe_2$ [2.142] show a small modulus dip ($\lesssim 0.1\%$) and attenuation maximum at T_i. Attempts to fit the combined frequency and temperature dependences to a single relaxation theory have proven unsuccessful. Thus far no Brillouin scattering or optical studies of the central peak have been carried out in this system. Nevertheless 2H-$TaSe_2$ is perhaps the best understood incommensurate system from the microscopic point of view, due to the well-established electron-phonon coupling mechanism. The reader is referred to the literature [2.17] for further details.

2.8.2 Incommensurate-Ferroelectric: K_2SeO_4

Incommensurate transitions need not involve free carriers. An active competition among forces of differing range is often sufficient to produce a lattice instability at an incommensurate wave vector [2.17]. A particularly beautiful and well-studied example is provided by K_2SeO_4 which exhibits an incommensurate $[\underline{q}_i = (a_c^*/3)(1-\delta)$, $a^* = 2\pi/a]$ second-order transition at $T_i = 130$ K followed by a first-order lock-in transition at $T_{com} = 93$ K to a ferroelectric cell three times as long in the $\hat{a}$ direction as the original cell above T_i. Neutron scattering dispersion curves [2.133] shown in Fig.2.22 reveal a striking softening of the $\hat{a}$ directed Σ_2 phonon displayed in an extended zone scheme. In addition, close to T_i an unresolved singular central peak is seen. The inelastic behavior near T_i is characterized by [2.133]

$$[\hbar\omega(\underline{q})]^2 = [\hbar\omega(q_i)]^2 + \beta_1'(q_x - q_i)^2 + \beta_2'q_y^2 + \beta_3'q_z^2 \quad , \tag{2.37}$$

where $\beta_1' = 110$ meV^{2}Å^2; $\beta_2' = 570$ meV^{2}Å^2; and $[\hbar\,(\underline{q}_i)]^2 = A(T-T_i)$ with $A \approx 0.065$ meV^2/K. In the mean field approximation and following the simple scenario sketched above, we would expect that immediately below T_i the amplitude mode at $\underline{q} = 0$ would have a frequency $\omega_A^2 = 2A(T_i - T)$ and the phase mode would have $\hbar\omega_{Ph}^2(\underline{q}) = \beta_1'(q_x)^2 + \beta_2'q_y^2 + \beta_3'q_z^2$.

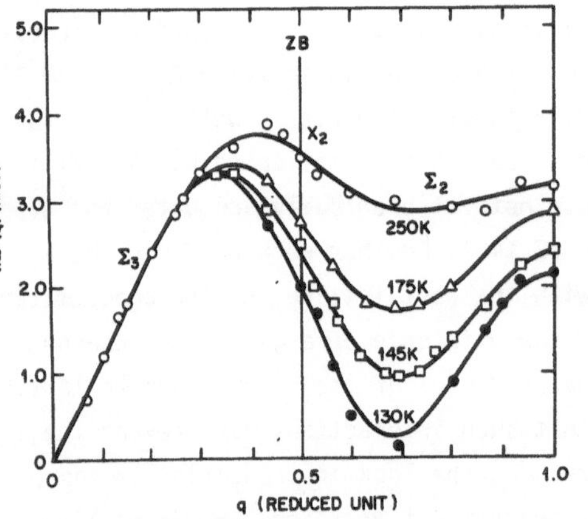

Fig. 2.22. Temperature dependence of Σ_3 branch phonon dispersion curve in K_2SeO_4, determined from inelastic neutron scattering. Softening at the incommensurate wave vector $q^* \approx 0.7$ is evident as T is reduced toward $T_i = 129$ K (after [2.133])

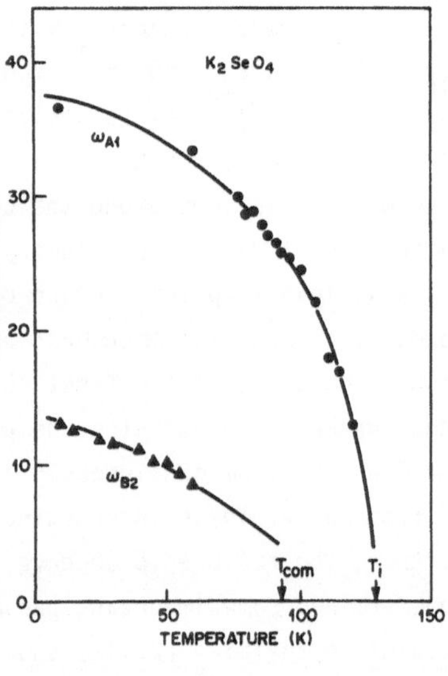

Fig. 2.23. Soft mode frequencies for K_2SeO_4 below T_i as determined by Raman scattering [2.146]

The original Raman studies of K_2SeO_4 [2.143,144] were directed at understanding the ferroelectric commensurate phase below T_{com} and were insufficiently precise to probe either of these modes. A more careful study by WADA et al. [2.145,146] found a soft optic mode of A_1 symmetry whose temperature behavior suggested it be identified with the amplitude mode. As shown in Fig.2.23, ω_A could be followed up to within ≈ 20 K of T_i. More recent experiments [2.146] in the commensurate phase revealed the lowest B_2 mode at 12.8 cm^{-1} evolving up to 67 K. This mode appears to soften toward T_{com} rather than T_i and has therefore been identified as the *precursor* to the phason in the incommensurate phase. No observation of the *phason* itself, however, could be made. Further, no Brillouin, ultrasonic, or dynamic central peak studies have been reported for K_2SeO_4, but they are clearly called for. Thus, the situation in K_2SeO_4 is quite similar to that in 2H-$TaSe_2$ insofar as experimental studies of elementary excitations are concerned. The nature of the distortions appears simpler in K_2SeO_4 and has been thoroughly explored by IIZUMI et al. [2.133], but no microscopic theory, such as exists for $TaSe_2$, has been developed.

2.8.3 Ferroelectric-Antiferroelectric: $BaMnF_4$

$BaMnF_4$ is perhaps the most complicated and least understood of the incommensurates [2.147]. At room temperature it is orthorhombic C_{2v}^{12} and ferroelectric. Below 26 K the Mn^{2+} ions order antiferromagnetically and a linear magnetoelectric effect exists. This latter behavior together with the original temperature-dependent Raman studies led RYAN and SCOTT [2.148] to suggest that $BaMnF_4$ undergoes a cell doubling phase transition near 250 K driven by a soft zone boundary phonon at $\underline{q}_i$ = (0, 0.5, 0.5). Elastic neutron studies by COX et al. [2.149] revealed that while a cell-multiplying transition does indeed occur, the low-temperature phase is incommensurate and the critical wave vector is $\underline{q}_i$ = (0.392, 0.5, 0.5). Additional superlattice reflections at $\underline{q}_i'$ = (0.216, 0, 0) and $\underline{q}_i''$ = (0.176, 0.5, 0.5) could be understood as arising from second and third harmonics of the primary order parameter at $\underline{q}_i$. More recent inelastic neutron studies [2.150] above T_i = 247 K show an overdamped soft mode whose width narrows upon approach to T_i as well as a diverging but unresolved central peak. No phasons have been observed in either the neutron or the Raman experiments.

Inasmuch as $BaMnF_4$ exhibits no lock-in transition - indeed δ is apparently temperature independent - the soft mode observed by RYAN and SCOTT [2.148] to decrease in frequency from ≈ 40 cm^{-1} upon heating toward T_i (coupling to the 31 cm^{-1} B_2 mode along the way) is probably the amplitude mode. Again the phason has not been observed. Some possible indirect evidence for it may have been found in the acoustic behavior, however. FRITZ [2.151] discovered a large (25%) velocity minimum for 4-30 MHz) transverse acoustic waves propagating parallel to c, polarized along b. A smaller (5%) dip superimposed on a 5% step velocity decrease was also found for c-directed longitudinal waves. Figure 2.24 shows this behavior. In addition, critical

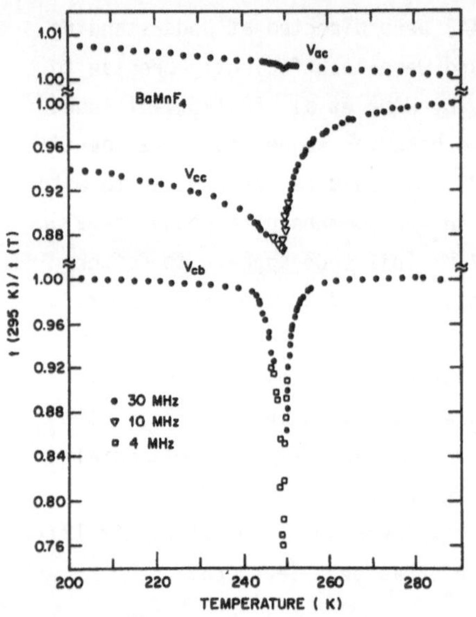

Fig. 2.24. Velocity anomalies accompanying the incommensurate structural transition at 250 K in $BaMnF_4$. Time-of-flight ultrasonic pulse technique was used (after [2.151]) [2.151])

acoustic attenuation was found which exhibited unusual frequency dependence [2.152] However, Brillouin scattering experiments of BECHTLE and SCOTT [2.153] on the transverse c phonons revealed that high-frequency (9 GHz) phonons were unaffected by the phase transition, suggesting a strong relaxation near T_i. Detailed Brillouin results obtained at smaller scattering angles could be fitted to a temperature-dependent relaxation frequency of the form $\omega_R = \omega_0[(T_i - T)/T_i] + \omega_1$ where $\omega_0 = 2\times10^{12}$ Hz; $\omega_1 = 4.2\times10^9$ Hz. They speculated that the origin of this acoustic relaxation might lie in an overdamped phason. However, in view of similar effects in commensurate transitions like NH_4Cl [2.154] and in the persistence of the relaxation effects above T_i, where the phason does not exist, this speculation remains unsupported. Clearly a direct observation of the phason below T_i is needed here as well.

In concluding this section on incommensurate phase transitions, we remark that much remains to be done regarding the critical dynamics of even the most thoroughly studied incommensurates. Progress in this area depends crucially upon finding a system and a technique that will probe the phason quantitatively and more or less directly. There is reason for optimism in that several new incommensurate transitions are sure to be discovered in the near future, and some will prove amenable to this type of inquiry. In view of the expected characteristic frequencies of the phason modes, it seems likely that light scattering will play a role.

2.9 Nonlinear Optical Studies of Phase Transitions

In all of the preceding discussion we have assumed that a linear approximation to the dielectric response of a crystal was sufficient, and that P could be related to E by a tensor of second rank. That is,

$$\underline{P} = \underline{\chi}^{L} \cdot \underline{E} \quad . \tag{2.38}$$

Note that $\underline{\chi}^{L}$ is the electrical susceptibility at *optical* frequencies, and should not be confused with the static susceptibility $\chi = \langle\psi\psi\rangle$ which has been used in previous sections. The anomalies in $\underline{\chi}^{L}$ observed at phase transitions are typically small in magnitude. On the other hand, the nonlinear terms, omitted from (2.38) often exhibit order of magnitude changes at phase transitions, and, therefore, afford useful experimental probes [2.14,155]. An informative review of the application of such experimental techniques to phase transitions has been given by VOGT [2.14].

Two assumptions underlie (2.38). These are I) superposition of plane waves and II) the electric dipole approximation. In the latter the variation of the electric field with position, even at optical wavelengths, is ignored. Relaxing these two restrictions gives, using the notation of VOGT [2.14],

$$\underline{P} = \underline{\chi}^{L} \cdot \underline{E} + \underline{\chi}^{A} : \nabla\underline{E} + \underline{\chi}^{NL} : \underline{E}\underline{E} + \underline{\chi}^{N} \vdots \underline{E}\underline{E}\underline{E} + \dots \quad . \tag{2.39}$$

The tensors of rank higher than two are responsible for such phenomena as optical activity ($\underline{\chi}^{A}$), and second harmonic generation ($\underline{\chi}^{NL}$, $\underline{\chi}^{N}$), as well as second harmonic scattering, two-photon absorption, etc. We shall not consider these latter phenomena here because they have seen little application to the study of phase transitions [2.156,157].

Nonlinear optical processes have been applied to the study of molecular mechanics at phase transitions by BERGMAN [2.158,159]. He assumed that it is possible to express the nonlinear susceptibility of the crystal in terms of the corresponding nonlinear susceptibilities of individual bonds. The temperature dependence of second harmonic generation (SHG), for example, is assumed to result only from the changing orientation of these bonds, rather than from any change in their individual susceptibilities. By the use of appropriate geometrical considerations, considerable information can be derived on the basis of this model regarding the evolution of structure near phase transitions.

Consider a given bond to be characterized by a nonlinear susceptibility $\beta_{\ell mn}$. The contribution of this bond to the macroscopic quantity $\underline{\chi}^{NL}$ is given by the relation

$$\underline{\chi}^{NL}_{ijk} = V^{-1} G_{i\ell} G_{jm} G_{kn} \beta_{\ell mn} \quad , \tag{2.40}$$

where V is the unit cell volume, and $\underline{G}$ is the transformation tensor taking the coordinates of the individual bond into the crystal coordinates. In (2.40) the summation over repeated indices is implied. The observed value of $\underline{\chi}^{NL}_{ijk}$ is the result of summing (2.40) over all bonds in the unit cell. This equation is considerably simplified if the individual bonds are assumed to possess $C_{\infty v}$ symmetry. Then, for example, a diagonal member of $\underline{\chi}^{NL}$ is given by

$$\underline{\chi}^{NL}_{333} = V^{-1} \sum_i [m_i^3 \beta_i^{\parallel} + 3m_i(1 - m_i^2)\beta_i^{\perp}] \quad , \tag{2.41}$$

where m_i represents the direction cosine of the bond axis with respect to the crystal z axis, and the summation extends over all bonds in a unit cell. The tensor β now has only two independent components, $\beta_{333} \equiv \beta^{\parallel}$ and $\beta_{311} = \beta_{113}$ = etc. $\equiv \beta^{\perp}$. Further simplification can stem from known properties of the structure of a given substance. For example in $Pb_5Ge_3O_{11}$ the symmetry of the Ge_2O_7 group nearly prevents its contributing to $\underline{\chi}^{NL}$. Hence in this case $\underline{\chi}^{NL}$ reflects mainly the rotation of the GeO_4 tetrahedron. From this model and the known high-temperature structure it is possible to extract from SHG measurements the rotation angle [2.155] and thus infer the order parameter. The method works very well and gives results which show good internal consistency as well as agreement with known structural results, which are often less precise. In addition to $Pb_5Ge_3O_{11}$ the method has been applied to transitions in SiO_2 [2.155], $LiNbO_3$ [2.158], $LiTaO_3$ [2.160], $PbTiO_3$ [2.161], and $Ba_6Ti_2Nb_8O_{30}$ [2.159]. Furthermore, in several cases (e.g., [2.158]) it has been possible to calculate the behavior of the spontaneous polarization P_s from the same model and to show good agreement with experimental results. It should be pointed out that the method is *not* suited to *determination* of complicated structures. However, once the basic structural modification is known from other techniques (e.g., X-ray diffraction), SHG may afford an easy and readily interpreted probe of the temperature dependence of that structural modification. Another important limitation of this technique may result from the uncontrolled effect of domains on phase matching conditions, which in many cases could prevent extraction of quantitative information.

All the cases mentioned above rest upon the assumption that the average value of β may be used in (2.40). That is, we assume $\langle\beta:\underline{F}\underline{F}\rangle = \langle\beta\rangle : \langle\underline{F}\rangle\langle\underline{F}\rangle$, where $\underline{F}$ is the local field. This mean field treatment works well in the displacive phase transitions mentioned above. A very interesting breakdown of this assumption occurs in $NaNO_2$ [2.162]. $NaNO_2$ undergoes a pair of phase transitions in which the NO_2^- ions order orientationally, at the same time driving a displacement of the Na^+ ions, resulting in ferroelectricity in the lowest phase. In this case, as opposed to those above, the behaviors of $\underline{\chi}^{NL}$ and P_s are very different from each other. This can be traced to their microscopic origin. The value of P_s results from the displacement of the Na ions, while $\underline{\chi}^{NL}$ is mainly affected by the orientation of the NO_2- ions. Also the breakdown of the mean field assumption is dramatically illustrated by the existence of SHG in the high-temperature phase which, on average, possesses inversion symmetry, which should preclude SHG. The essential reason for this is that, while $\langle\beta\rangle = 0$ for a given site, existence of *short range order* above T_c means that $\langle\beta:\underline{\underline{EE}}\rangle \neq 0$. Mathematically, this is expressed as the last term in (2.39). Essentially then, SHG becomes a probe of short range order in the macrosopically disordered phase. The experimental results, shown in Fig.2.25, are not yet fully understood, especially in light of the recent evidence for incommensurate character of the intermediate phase in $NaNO_2$, but the theoretical challenge they pose is one that will hopefully be met in the near future.

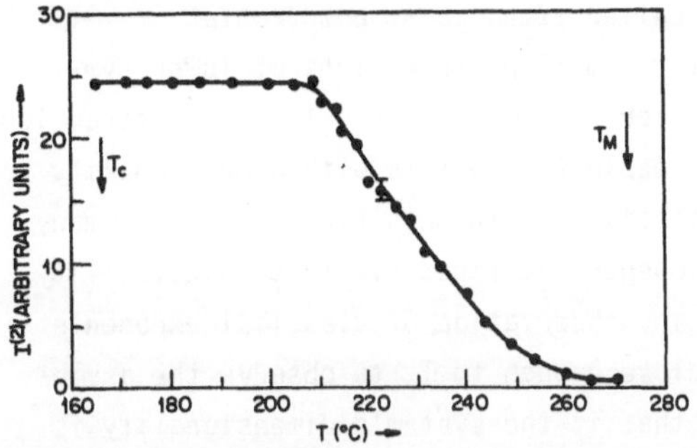

Fig. 2.25. The temperature dependence of the SHG intensity in the disordered phase of $NaNO_2$ between the transition temperature T_c and the melting temperature T_m, indicating the effect of short range order

In summary then nonlinear optical techniques are capable of probing aspects of phase transitions which are not accessible by linear techniques. Application of this important class of techniques will probably see further growth in coming years. The possibility of studying dynamic short range order (i.e., order parameter fluctuations) above T_c via SHG has, to our knowledge, received little attention. This possibility represents an unexplored but attractive area.

2.10 Renormalization Group Theory and Multicritical Phenomena

Most of the refinements in the theory of critical phenomena growing out of the Wilson renormalization group theory (RGT) have been applied to magnetic and superfluid transitions [2.44,163]. Nevertheless, some of their implications for structural transitions deserve a brief discussion here. With few exceptions, optical experiments on structural transitions have been interpreted within the framework of mean field theory. However, as experiments probe closer to T_c and as dependences on other external variables such as electric field, pressure, uniaxial stress, etc., are increasingly investigated near structural transitions, a number of additional considerations become important. These include an awareness of a) RGT predictions for the class of transition under study and b) possible fundamental complications such as the change of universality class associated with a particular external perturbation. There are sufficiently few entries in each of these categories that we can mention most of them.

The general approach in RGT begins with a classification of the dimensionality of the relevant space and order parameters. Then the Hamiltonians are classified according to the symmetries and powers of the order parameters they contain [2.163]. Hamiltonians with similar symmetries are said to belong to the same universality class. The Hamiltonians are examined under successive applications of the renormalization operator R so as to determine the so-called *fixed point Hamiltonian* ($RH^* = H^*$). The stability of H^* with respect to small perturbations of lower symmetry is also examined. It is important that for systems with short range interaction the so-called Gaussian fixed point is never stable for systems with dimensionality less than four. This is important because associated with this fixed point are many of the "classical" (mean field) critical exponents. At first sight, the instability of the Gaussian fixed point would mean that any observation of classical exponents implies that the experiment has not probed close enough to T_c to observe the asymptotic behavior. It should be noted however, that if the system's dimensionality is equal to the "marginal dimensionality" for the relevant universality class only logarithmic departures from mean field theory are expected and may be exceedingly difficult to detect experimentally [2.164]. COWLEY [2.165] has pointed out, moreover, that there are situations in which even RGT predicts mean field behavior for real materials. These situations involve structural phase transitions in which a homogeneous deformation of the crystal's unit cell is a primary order parameter. For example, whenever the soft mode is Raman active on both sides of the phase transition, the resulting linear coupling with acoustic phonons will make strain a primary order parameter. Perhaps the most familiar examples are ferroelectric materials such as KDP which are piezoelectric in their paraelectric phases. Also included here are cooperative Jahn-Teller systems such as $PrAlO_3$. The rare earth vanadates RVO_4 (R = Tb, Tm, Dy, ...) also exhibit cooperative Jahn-Teller structural transitions [2.7]. Birefringence and other order parameter measurements yield classical values of $\beta = \frac{1}{2}$ and $\gamma = 1$ for all members of this family, except for $DyVO_4$. There the measured β is closer to 1/3, presenting a presently unresolved puzzle [2.166]. With this one exception, the experiments to date bear out Cowley's observation that the result of coupling to acoustic waves is to decrease the effects of fluctuations and to suppress by one the dimensionality which separates classical from nonclassical behavior.

Transitions in which no linear acoustic coupling exists above T_c are not covered by these arguments, so that classical behavior is not expected for nonpiezoelectric-ferroelectrics like $Pb_5Ge_3O_{11}$ or for cell-multiplying transitions like $SrTiO_3$. The latter class of transition has been considered within RGT [2.167,168]. COWLEY and BRUCE [2.167] constructed an effective Hamiltonian for $SrTiO_3$ and after examination of its stability concluded that the cubic-tetragonal transition should exhibit the same static exponents as the classical isotropic Heisenberg model, specifically $\beta = 0.373$. This value for β is somewhat higher than the 0.33 reported from EPR [2.107]. Subsequently it was shown [2.168] that in anisotropically stressed perovskites critical behavior appropriate to either the Ising, the x-y, or the Heisenberg models may be realized, depending on the size and symmetry of applied stresses. This work points out that close to T_c residual stresses (such as might arise from preparation of a monodomain sample, for instance) might change the nature of the fixed point. Even more severe might be the effects of deliberately imposed external stress (as has been applied [2.169] in the study of the cubic-tetragonal-trigonal evolution in $SrTiO_3$). Figure 2.26 shows a calculated phase diagram for $SrTiO_3$ subject to stress directed along [100]. For $T > T_b$ and for positive stress, the transition is characterized by x-y behavior, while for negative stress Ising behavior is found. These phase transition lines meet at T_b, a so-called *bicritical point* at zero stress. For $T < T_b$ there is a line of first order "spin-flop-like" phase transitions. Depending upon the values of certain Hamiltonian coefficients, other stresses may lead to even more complicated multicritical phenomena. Although these arguments appear to account for the EPR measurements of β in $SrTiO_3$, there have to date been no optical or light scattering experiments which have examined these predictions in detail.

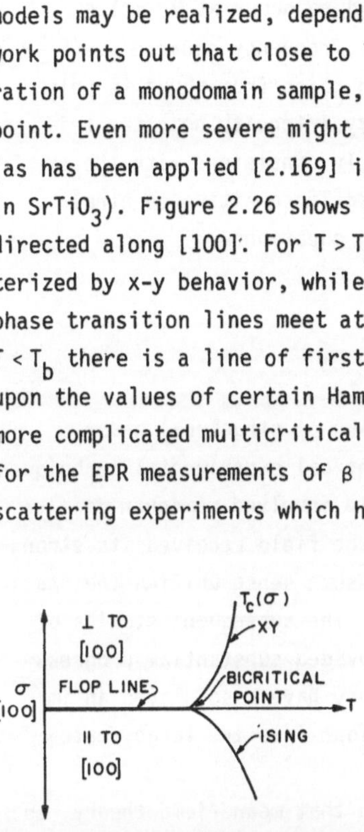

Fig. 2.26. Phase diagram of a perovskite under [100] applied stress σ. For $\sigma < 0$ the pseudospin order parameter is parallel to [100] and Ising exponents are predicted. For $\sigma > 0$, the order is perpendicular to [100] and xy exponents are predicted. At $\sigma = 0$, a first-order spin-flop-like transition occurs for $T < T_b$

Indeed only two light scattering investigations [2.170,171] of multicritical point structural transitions have been reported. These have provided semiquantitative information on the use of pressure to access a tricritical point. A tricritical point occurs where a line of second-order transitions joins smoothly onto a line of first-order transitions. First, we consider the order-disorder transition in NH_4Cl. At low pressures this transition is first order, but apparently becomes second order for $P \geq 1.3$ kbar. No soft mode has ever been seen for this transition at any pressure.

But FRITZ and CUMMINS [2.170] have concluded from measurements of the unresolved Rayleigh scattering that the NH_4Cl tricritical point occurs at $P_i = 1.3$ kbar; $T_t = 255$ K, in agreement with earlier thermal expansivity studies.

The ferroelectric transition in SbSI, which occurs at 293 K and 1 bar, does exhibit a well-studied soft mode. Using pressure as an additional variable, PEERCY [2.171] has measured both the elastic scattered intensity and the ferroelectric soft mode behavior between 120 and 300 K. Both types of measurement suggest that for $P < 1.4$ kbar, the transition is first order, while for $P > 1.4$ kbar it becomes second order. The tricritical point is thereby determined to occur at $P_t = 1.40$ kbar and $T_t = 235$ K. Further, the slope of the soft mode frequency squared vs $(P - P_c)$ changes by nearly the factor of two expected (by MFT) in going from above to below T_t. Although the behavior of the integrated low-frequency intensity near the tricritical point observed in both materials is qualitatively consistent with the expectations of critical opalescence, the lack of sufficiently sensitive spectral resolution leaves unanswered some of the most intriguing questions.

2.11 Concluding Remarks

In this review we have seen the evolution of understanding of structural phase transitions through the viewpoint provided by various optical experimental techniques. Although the earliest application of optics in this field involved semiquantitative determination of order parameters using birefringence, the field received its strongest stimulus from the realization that the soft mode in some sense unifies the static and the dynamic aspects of structural phase transitions. The subsequent studies of soft modes using both scattering and absorption have provided substantial progress in the microscopic understanding of such transitions. They have also played an important role in the incorporation of structural transitions into the large framework of critical phenomena.

This incorporation in turn has led to the realization that mean field theory, so often successful in describing experimental results on structural transitions, should not apply to them all. Indeed an increasing amount of the experimental activity regarding structural transitions has been concerned with exploring departures from MFT. The departures from mean field theory which have been explored by optical techniques thus far have been few in number. The fluctuation contributions to the birefringence in strontium titanate and the integrated scattered intensity in lead germanate are two salient examples. However, we should mention that many unexplored possibilities in this area afforded by the techniques of nonlinear optics deserve increased consideration from both theorists and experimentalists.

Optical experiments have also played an important role in explicating other breakdowns in and refinements of the early quasi-harmonic soft mode ideas. In particular the complications associated with interactions of the soft mode with other degrees of freedom have by now been thoroughly explored using the techniques of light scattering. Other advances in the light scattering technique have permitted the first quantitative information to be obtained regarding the as-yet incompletely understood "central peak" phenomenon. Perhaps the most important result to date in this area is the verification that the central peaks may have both static and dynamic contributions, and that therefore no single mechanism can account for the observations. Evidence is strong that both scattering from defects and the effects of soft mode coupling to phonon density fluctuations contribute to the central peaks studied in both strontium titanate and lead germanate.

Other important directions for future study include attempts to observe directly the phason dynamics associated with incommensurate phase transitions, and more complete explorations of multicritical points using pressure, stress, and other external fields as additional experimental variables.

An underlying theme of this review has been the powerful specificity which optical techniques afford the study of particular aspects of structural transitions. When compared to other techniques the rich and useful array of selection rules accompanying optical experiments permits separation of much otherwise confusing information. This fact has played a determining role in the way in which we have structured this review. That is, the relatively few examples which we have chosen to discuss in detail have been selected to represent each of the different symmetry classes important from an optical point of view. While we are fairly certain that the choices we have made are satisfactory from that point of view, we are equally certain that different authors in constructing a review on this topic might have chosen other examples. Thus it is our hope that the extended bibliography which comprises the final section of this paper will partially rectify this difficulty.

References

2.1 M. Balkanski, R.C.C. Leite, S.P.S. Porto (eds.): *Proceedings of the 3rd International Conference on Light Scattering in Solids* (Flammarion, Paris 1976)
2.2 K.A. Müller, A. Rigamonti (eds.): *Proceedings of the International School of Physics 'Enrico Fermi', Course LIX. Local Properties at Phase Transitions* (North-Holland, Amsterdam 1976)
2.3 E.J. Samuelson, E. Anderson, J. Feder (eds.): *Structural Phase Transitions and Soft Modes* (Universitetsforlaget, Oslo 1971)
2.4 Y. Farge: Nuovo Cimento *39*, 356 (1977)
2.5 I.R. Jahn, H. Dachs: Fortschr. Mineral. *51*, 176-239 (1974)
2.6 E. Courtens: In Ref. 2.2, pp.293-311
2.7 G.A. Gehring, K.A. Gehring: Rep. Prog. Phys. *38*, 1-89 (1975)
2.8 G.A. Smolensky, Z.M. Hashkozev: Mater. Res. Bull. *6(10)*, 1065-1074 (1971)
2.9 P.S. Peercy: In Ref. 2.1, pp.782-95
2.10 P.A. Fleury: In Ref. 2.1, pp.747-60
2.11 J.F. Scott: Rev. Mod. Phys. *46*, 83-128 (1974)
2.12 E.F. Steigmeier: Ferroelectrics *7*, 65-70 (1974)

2.13 T. Nakamura: Ferroelectrics *9*, 159-169 (1975)
2.14 H. Vogt: Appl. Phys. *5*, 85-96 (1974)
2.15 R.A. Cowley: Ferroelectrics *6*, 163 (1974)
2.16 H. Granicher, K.A. Muller: Mater. Res. Bull. *6*, 977-988 (1971)
2.17 T. Riste (ed.): *Electron-Phonon Interactions and Phase Transitions* (Plenum, New York 1978)
2.18 H. Thomas: In Ref. 2.2, pp.3-44
2.19 A.D. Bruce: Ferroelectrics *12*, 21-29 (1976)
2.20 A.I. Larkin, D.E. Khmel'nitskii: Sov. Phys.-JETP *29*, 1123 (1969)
2.21 T.S. Robinson, W.C. Price: Proc. Phys. Soc. *B66*, 969-974 (1953)
2.22 J.J. Chamberlain, E. Gibbs, H.A. Gebbie: Nature *198*, 874 (1963)
2.23 D.A. Ledsham, W.G. Chambers, T.J. Parker: Infrared Phys. *16*, 515-522 (1976)
2.24 D.A. Ledsham, W.G. Chambers, T.J. Parker: Infrared Phys. *17*, 165-172 (1977)
2.25 T.J. Parker, W.G. Chambers: Infrared Phys. *14*, 207-215 (1974)
2.26 P.R. Staal, J.E. Eldridge: Infrared Phys. *17*, 299-303 (1977)
2.27 P.R. Griffiths: *Chemical Infrared Fourier Transform Spectroscopy* (Wiley, New York 1975)
2.28 M.F. Kimmitt: Infrared Phys. *17*, 459-466 (1977)
2.29 V.G. Bochkow, V.I. Bugakov, K.A. Verkhovskaya, T.M. Polkhovskaya, V.M. Fridkin: Sov. Phys.-Solid State *16*, 1217-1219 (1975)
2.30 V.M. Fridkin, K.A. Verkovskaya, B.G. Bochkov: Phys. Status Solidi A*22*, 759-766
2.31 A. Abragam, B. Bleaney: *Electron Paramagnetic Resonance of Transition Ions* (Clarendon, Oxford 1970) Chap.16
2.32 F. Jona, G. Shirane: *Ferroelectric Crystals* (Macmillan, New York 1962)
2.33 R.T. Harley, W. Hayes, A.M. Perry, S.R.P. Smith: Solid State Commun. *14*, 521-524 (1974)
2.34 M. Born, E. Wolf: *Principles of Optics*, 5th ed. (Pergamon, New York 1975)
2.35 A.R. Johnston, J.M. Weingart: J. Opt. Soc. Am. *55*, 828 (1965)
2.36 E.H. Izen, F.A. Modine: Rev. Sci. Instrum. *43*, 1563 (1972)
2.37 F.A. Modine, R.W. Major, E. Sonder: Appl. Opt. *14*, 757 (1975)
2.38 G.A. Gehring: J. Phys. C*10*, 531-542 (1977)
2.39 A.R. Johnston: Appl. Opt. *8*, 1837-1848 (1969)
2.40 F. Micheron, G. Bismuth: Rev. Sci. Instrum. *43*, 292 (1972)
2.41 P.A. Fleury: J. Acoust. Soc. Am. *49*, 1041 (1971)
2.42 J.B. Lastovka: Bell Syst. Tech. J. *55*, 1225 (1976)
2.43 P.A. Fleury: Comments Solid State Phys. *4*, 149,167 (1972)
2.44 P.C. Hohenberg, B.I. Halperin: Rev. Mod. Phys. *49*, 435-475 (1977)
2.45 S.I. Mizushima: In *Light and Matter II, Encyclopedia of Physics*, Vol. XXVI, ed. by S. Flügge (Springer, Berlin, Heidelberg, New York 1971) pp.171-243
2.46 I.L. Birman: In *Symposium on Ferroelectricity*, ed. by E.F. Weller (Elsevier, Amsterdam 1967) p.20, et seq.
2.47 J.M. Worlock: In Ref. 2.3, pp.329-332
2.48 M. Singh: Indien J. Pure Appl. Phys. *15*, 338 (1977)
2.49 C.H.F. Velzel: J. Opt. Soc. Am. *67*, 1021-1027 (1977)
2.50 K.B. Lyons, P.A. Fleury: J. Appl. Phys. *47*, 4898-4900 (1976)
2.51 D.S. Cannel, G.B. Benedek: Phys. Rev. Lett. *25*, 1157 (1970)
2.52 J.R. Sandercock: RCA Rev. *36*, 89-107 (1975)
2.53 J. Sandercock: Bull. Am. Phys. Soc. *23*, 387-388 (1978)
2.54 K.H. Langley, N.C. Ford: J. Opt. Soc. Am. *59*, 281-284 (1969)
2.55 W. Cochran: Adv. Phys. *9*, 387 (1960)
2.56 J.C. Toledano: Ann. Telecommun. *29*, 1 (1974)
2.57 K. Aizu: J. Phys. Soc. Jpn. *33*, 629-634 (1972)
2.58 A. Sawada, M. Udagawa, T. Nakamure: Phys. Rev. Lett. *39*, 829-832,902 (1977)
2.59 R.J. Eliott, R.T. Harley, W. Hayes, S.R.P. Smith: Proc. R. Soc. London A*328*, 217-266 (1972)
2.60 E. Pytte: Ferroelectrics *7*, 193 (1974)
2.61 R.T. Harley, W. Hayes, A.M. Perry, S.R.P. Smith: J. Phys. C*6*, 2382-2400 (1973)
2.62 R.T. Harley, W. Hayes, A.M. Perry, S.R.P. Smith: J. Phys. C*8*, L123-125 (1975)
2.63 K.B. Lyons, R.J. Birgeneau, E.I. Blount, L.G. van Uitert: Phys. Rev. B*11*, 891-900 (1975)
2.64 M.D. Sturge, E. Cohen, L.G. van Uitert, R.P. van Stapele: Phys. Rev. B*11*, 4768-4779 (1975)

2.65 P.A. Fleury, P.D. Lazay, L.D. van Uitert: Phys. Rev. Lett. *33*, 492-495 (1974)
2.66 P.A. Fleury, J.M. Worlock: Phys. Rev. *174*, 613-623 (1968)
2.67 K.B. Lyons, P.A. Fleury: Phys. Rev. B *17*, 2403 (1978)
2.68 K.H. Germann, W. Muller-Lierheim, H.H. Otto, T. Suski: Phys. Status Solidi A*35*, K165-167 (1976)
2.69 W. Muller-Lierheim, H.H. Otto: Solid State Commun. *24*, 349-352 (1977)
2.70 H. Iwasaki, S. Miyazawa, H. Koizumi, K. Sugii, N. Niizeki: J. Appl. Phys. *43*, 4907-4915 (1972)
2.71 W. Muller-Lierheim, T. Suski, H.H. Otto: Phys. Status Solidi B*80*, 31-41 (1977)
2.72 K. Hisano, J.F. Ryan: Solid State Commun. *11*, 1745-1749 (1972)
2.73 W. Taylor, D.J. Lockwood, J.W. Arthur, T.J. Hosea: Ferroelectrics *12*, 113-115 (1976)
2.74 P.A. Fleury, K.B. Lyons: Phys. Rev. Lett. *37*, 1088-1091 (1976)
2.75 R.A. Cowley, G.J. Coombs: J. Phys. C*6*, 143 (1973)
2.76 D.J. Lockwood, J.W. Arthur, W. Taylor, T.J. Hosea: Solid State Commun. *20*, 703-707 (1976)
2.77 W. Cochran: Adv. Phys. *10*, 401 (1961)
2.78 I.P. Kaminow, T.C. Damen: Phys. Rev. Lett. *20*, 1105 (1968)
2.79 R.S. Katiyar, J.F. Ryan, J.F. Scott: Phys. Rev. B*4*, 2635 (1971)
2.80 N. Lagakos, H.Z. Cummins: Phys. Rev. B*10*, 1063-1096 (1974)
2.81 N. Lagakos, H.Z. Cummins: Phys. Rev. Lett. *34*, 883-886 (1975)
2.82 R.C. Leung, N.E. Tornberg, R.P. Lowndes: J. Phys. C*9*, 4477-4490 (1976)
2.83 R.P. Lowndes, N.C. Tornberg, R.C. Leung: Phys. Rev. B*10*, 911 (1974)
2.84 E.M. Brody, H.Z. Cummins: Phys. Rev. B*9*, 179-196 (1974)
2.85 J. Azoulay, D. Gerlich, E. Wiener, I. Pelah: Ferroelectrics *8*, 599-602 (1974)
2.86 J. Azoulay, D. Gerlich, E. Wiener-Avnear, I. Pleah: Phys. Status Solidi B*81*, 295-306 (1977)
2.87 I.A. Iakovlev, T.S. Velichkina, L.F. Mikheeva: Sov. Phys.-Dokl. *1*, 215 (1956)
2.88 T. Riste, E.J. Samuelson, K. Olson, J. Feder: Solid State Commun. *9*, 1455 (1971)
2.89 S.M. Shapiro, H.Z. Cummins: Phys. Rev. Lett. *21*, 1578 (1968)
2.90 L.N. Durvasula, R.W. Gammon: Phys. Rev. Lett. *38*, 1081-1048 (1977)
2.91 E. Courtens: Phys. Rev. Lett. *39*, 561-564 (1977)
2.91a E. Courtens: Phys. Rev. Lett. *41*, 1171 (1978)
2.92 M.D. Mermelstein, H.Z. Cummins: Phys. Rev. B*16*, 2177-2183 (1977)
2.93 K.B. Lyons, P.A. Fleury: Phys. Rev. Lett. *37*, 161-164 (1976)
2.94 P. Bastie, M. Vallade, C. Vettier, C.M.E. Zeyen: Phys. Rev. Lett. *40*, 337-340 (1978)
2.95 V.H. Schmidt: Phys. Rev. Lett. *37*, 839-842 (1976)
2.96 E. Pytte, J. Fedder: Phys. Rev. *187*, 1077 (1969)
2.97 H. Thomas, K.A. Muller: Phys. Rev. Lett. *21*, 1256 (1968)
2.98 M. Rousseau, J.Y. Gesland, J. Jullieard, J. Nouet, J. Zaremswitch, A. Zaremswitch: Phys. Rev. B*12*, 1579 (1975)
2.99 B.H. Torrie and D.J. Lockwood: Ferroelectrics *8*, 583-584 (1974)
2.100 S. Hirotsu, S. Sawada: Solid State Commun. *12*, 1003-1005 (1973)
2.101 J. Berger, G. Hauret, M. Rousseau: Solid State Commun. *25*, 569-571 (1978)
2.102 E. Pytte: In Ref. 2.3, pp.171-188
2.103 K.B. Lyons, P.A. Fleury: Solid State Commun. *23*, 477-480 (1977)
2.104 K.S. Aleksandrov, A.T. Anistratov, A.I. Krupnyi, L.A. Pozdnyakova, S.V. Mel'nikova, B.V. Beznosikov: Sov. Phys.-Solid State *17*, 471-473 (1975)
2.105 F.A. Modine, E. Sonder, W.P. Unruh, C.B. Finch, R.D. Westbrook: Phys. Rev. B*10*, 1623-1634 (1974)
2.106 A.J. Rushworth, J.F. Ryan: Solid State Commun. *18*, 1239-1241 (1976)
2.107 K.A. Müller, W. Berlinger: Phys. Rev. Lett. *26*, 13-16 (1971)
2.108 E.F. Steigmeier, H. Auderset, G. Harbeke: In Ref. 2.3, pp.153-158
2.109 J.M. Worlock, D.H. Olson: In *Light Scattering in Solids*, ed. by M. Balkanski (Flammarion, Paris 1971) pp.410-414
2.110 T. Riste (ed.): *Anharmonic Lattices, Structural Transitions and Melting* (Noordhoff, Leiden 1974)
2.111 C.M. Varma: Phys. Rev. B*14*, 244-254 (1976)
2.112 B.I. Halperin, C.M. Varma: Phys. Rev. B*14*, 4030-4044 (1976)

2.113 T. Schneider, E. Stoll: In Ref. 2.2, pp.344-368
2.114 F. Schwabl: Ferroelectrics *7*, 395-396 (1974)
2.115 J.B. Hastings, S.M. Shapiro, B.C. Frazer: Phys. Rev. Lett. *40*, 237-240 (1978)
2.116 E. Courtens: Phys. Rev. Lett. *37*, 1584-1587 (1976)
2.117 L.E. Cross, A. Fouskova, S.E. Cummins: Phys. Rev. Lett. *21*, 812-814 (1968)
2.118 P.A. Fleury: Solid State Commun. *8*, 601-605 (1970)
2.119 J. Petzelt: Solid State Commun. *9*, 1485-1488 (1971)
2.120 E. Pytte: Solid State Commun. *8*, 2101-2104 (1970)
2.121 A.P. Levanyuk, D.G. Sannikov: Sov. Phys.-Solid State *12*, 2418 (1971)
2.122 B. Dorner, J.D. Axe, G. Shirane: Phys. Rev. B*6*, 1950-1963 (1972)
2.123 I.W. Shepherd: Solid State Commun. *9*, 1857-1860 (1971)
2.124 B.N. Ganguly, F.G. Ullman, R.D. Kirby, J.R. Hardy: Phys. Rev. B*12*, 3783-3788 (1975)
2.125 T. Shigenari, Y. Takagi, Y. Wakabayashi: Solid State Commun. *18*, 1271-1273 (1976)
2.126 V.T. Höchli: Phys. Rev. B*6*, 1814-1823 (1972)
2.127 Y. Lupin, G. Haaret: J. Phys. Lett. *35*, L-193 (1974)
2.128 J.M. Courdille, R. Deroche, J. Dumas: J. Phys. *36*, L-5 (1975)
2.129 Y. Luspin, G. Hauret: J. Phys. Lett. *35*, L193-195 (1974) (in French)
2.130 S. Itoh, T. Nakamura: Jpn. J. Appl. Phys. *14* (Suppl.) 183-186 (1975)
2.131 W.L. McMillan: Phys. Rev. B*12*, 1187-1199 (1975)
2.132 A.D. Bruce, R.A. Cowley: In *Proc. of Int'l. Conf. on Lattice Dynamics*, Paris, 1977 (Flammarion, Paris 1978)
2.133 M. Iizumi, J.D. Axe, G. Shirano, K. Shimaoka: Phys. Rev. B*15*, 4392-4411 (1977)
2.134 D.E. Moncton, J.D. Axe, F.J. DiSalvo: Phys. Rev. Lett. *34*, 734-737 (1975)
2.135 E.L. McMillan: Phys. Rev. B*14*, 1496-1502 (1976)
2.136 F.J. DiSalvo: In Ref. 2.17, pp.107-137
2.137 J.A. Holy, M.V. Klein, W.L. McMillan, S.F. Meyer: Phys. Rev. Lett. *37*, 1145-1148 (1976)
2.138 E.F. Steigmeier, G. Harbeke, H. Auderset, F.J. DiSalvo: Solid State Commun. *20*, 667-677 (1976)
2.139 J.C. Tsang, J.E. Smith, M.W. Shafer: Phys. Rev. Lett. *37*, 1407-1410 (1976)
2.140 W.L. McMillan: Phys. Rev. B*16*, 643-650 (1977)
2.141 W.L. McMillan: Phys. Rev. B*16*, 4655-4658 (1977)
2.142 M. Barmatz, L.R. Testardi, F.J. DiSalvo: Phys. Rev. B*13*, 4637-4639 (1976)
2.143 V. Fawcett, R.J.B. Hall, D.A. Long, V.N. Sankaranarayanan: J. Raman Spectrosc. *2*, 629-633 (1974)
2.144 V. Fawcett, R.J.B. Hall, D.A. Long, V.N. Sankaranarayanan: J. Raman Spectroc. *3*, 229-238 (1975)
2.145 M. Wada, A. Śawada, Y. Ishibashi, Y. Takagi: J. Phys. Soc. Jpn. *42*, 1229-1234 (1977)
2.146 M. Wada, H. Uwe, A. Sawada, Y. Ishibashi, Y. Tkagi. T. Sakudo: J. Phys. Soc. Jpn. *43*, 544-547 (1977)
2.147 J.F. Ryan, J.F. Scott: In Ref. 2.1, pp.761-769
2.148 J.F. Ryan, J.F. Scott: Solid State Commun. *14*, 5-9 (1974)
2.149 D.E. Cox, S.M. Shapiro, R.A. Cowley, M. Eibschutz, H.J. Guggenheim: Phys. Rev. B*19*, 5754 (1979)
2.150 S.M. Shapiro, D.E. Cox, R.A. Cowley, M. Eibschutz, H.J. Guggenheim: Bull. Am. Phys. Soc. *23*, 314 (1978)
2.151 I.J. Fritz: Phys. Lett. *51a*, 219-220 (1975)
2.152 I.J. Fritz: Phys. Rev. Lett. *35*, 1511-1514 (1975)
2.153 D.W. Bechtle, J.F. Scott: J. Phys. C*10*, L209-211 (1977)
2.154 P.D. Lazay, J.H. Luncacek, N.A. Clark, G.B. Benedek: In *Light Scattering Spectra of Solids*, ed. by G.B. Wright (Springer, Berlin, Heidelberg, New York 1969)
2.155 J.G. Bergman, G.R. Crane: Chem. Phys. Lett. *41*, 133-136 (1976)
2.156 I. Freund: Phys. Rev. Lett. *24*, 1017 (1970)
2.157 G. Dolino: Phys. Rev. B*6*, 4025 (1972)
2.158 J.G. Bergman: Chem. Phys. Lett. *38*, 230-233 (1976)

2.159 J.G. Bergman: Opt. Commun. *18*, 51-52 (1976)
2.160 J.G. Bergman: J. Am. Chem. Soc. *98*, 1054-1055 (1976)
2.161 J.G. Bergman, G.R. Crane, E.H. Turner: J. Solid State Chem. *21*, 127-133 (1977)
2.162 H. Vogt: Ferroelectrics *7*, 103-104 (1974)
2.163 M.E. Fisher: Rev. Mod. Phys. *46*, 597 (1974)
2.164 G. Ahlers, A. Kornblit, H.J. Guggenheim: Phys. Rev. Lett. *34*, 1656-1659 (1974)
2.165 R.A. Cowley: Phys. Rev. B*13*, 4877-4885 (1976)
2.166 R.T. Harley, R.M. Macfarlane: J. Phys. C*8*, L451-455 (1975)
2.167 R.A. Cowley, A.D. Bruce: J. Phys. C*6*, L191-196 (1973)
2.168 A. Aharony, A.D. Bruce: Phys. Rev. Lett. *33*, 427-430 (1974)
2.169 W.J.R. Burke, R.J. Pressley, J.C. Slonczewski: Solid State Commun. *9*, 121,191 (1971)
2.170 I.J. Fritz: H.Z. Cummins: Phys. Rev. Lett. *28*, 96-99 (1972)
2.171 P.S. Peercy: Phys. Rev. Lett. *35*, 1581-1584 (1975)

Bibliography

The accompanying bibliography is the result of an extensive literature search for the period between 1974 and the writing of this contribution. A somewhat less complete search of the 1971-1973 period is also included, as are references specifically cited in the text. Of the latter, those which do not fall into the categorization described below are listed as "uncategorized", at the end. Within all sections, references are listed alphabetically by first author. We have taken care to make this list as comprehensive as possible. Due to the tremendous amount of literature in this field we recognize that there probably exist omissions, perhaps significant ones. We apologize in advance to the authors for these inadvertent oversights. Titles are included in all cases to increase the usefulness to the reader. The first three sections apply to all phase transitions: Books and Reviews, Experimental Apparatus, and Theory. The remaining sections are broken up according to the translational symmetry of the order parameter (zone center, zone boundary, and incommensurate) and according to the optical techniques employed (birefringence, fluorescence and absorption), scattering and IR, and "other techniques"). It was decided to group scattering and infrared work together since they deal predominantly with the same phenomenon - namely optic phonons.

Naturally some articles fall into more than one category. In these cases the article is numbered only once and listed *without* a number in the other category(s). The choice of the "main" category in such cases was often arbitrary, but an effort was made to reflect the main thrust of the article in the choice.

A word is in order about the material in the categories involving incommensurate phase transitions. In a number of cases, materials previously thought to be of zone boundary type are now found to have incommensurate character. In such cases, *all* references on the materials in question, including $NaNO_2$, Thiourea, $(NH_4)_2BeF_4$, and $BaMnF_4$, have been placed under the heading of *incommensurate*, even though in many cases the articles themselves were not written from that viewpoint.

1 *Reviews and Books*

1 M. Balkanski (ed.): *Light Scattering in Solids* (Flammarion, Paris 1971)
2 M. Balkanski, R.C.C. Leite, S.P.S. Porto (eds.): *Proceedings of the 3rd International Conference on Light Scattering in Solids* (Flammarion, Paris 1976)
3 A.S. Barker: "Infrared Dielectric Behaviour of Ferroelectric Crystals", in *Ferroelectricity*, ed. by E.F. Weller (Elsevier, New York 1967) p. 213
4 J.L. Birman: "Symmetry Change, Phase Transitions, and Ferroelectricity", in *Symposium in Ferroelectricity*, ed. by E.F. Weller (Elsevier, Amsterdam 1967) p. 20, et seq.
5 R. Blinc, B. Zeks: *Soft Modes in Ferroelectrics and Antiferroelectrics* (North-Holland, Amsterdam 1974)
6 M. Born, E. Wolf: *Principles of Optics*, 5th ed. (Pergamon, New York 1975)
7 M. Cardona (ed.): *Light Scattering in Solids*, Topics in Applied Physics, Vol. 8 (Springer, Berlin, Heidelberg, New York 1975)
8 R. Claus, L. Merten, J. Brandmüller: *Light Scattering by Phonon-Polaritons*, Springer Tracts in Modern Physics, Vol. 75 (Springer, Berlin, Heidelberg, New York 1975)
9 R. Claus, J. Brandmüller: Aspects of application of IR reflectivity- and Raman scattering-experiments on polaritons in solid state chemistry and biophysics. Spec. Lett. *9*, 575-614 (1976)
10 R.A. Cowley: Structural phase transitions, soft modes, and quasielastic scattering. Ferroelectrics *6*, 163 (1974)
11 L.E. Cross (ed.): *Phase Transitions 1973* (Pergamon, New York 1973)
12 Y. Farge: Optical studies of phase transitions in solids. Nuovo Cimento *39*, 356 (1977)
13 M.E. Fisher: Renormalization group in the theory of critical behavior. Rev. Mod. Phys. *46*, 597 (1974)
14 P.A. Fleury: Phonon instabilities and interactions near solid state phase transitions. J. Acoust. Soc. Am. *49*, 1041 (1971)
15 P.A. Fleury: Soft mode phase transitions. Comments Solid State Phys. *4*, 149, 167 (1972)
16 P.A. Fleury: The effects of soft modes on the structure and properties of materials. Annu. Rev. Mater. Sci. *6* (1976) p. 157
17 P.A. Fleury: "Recent Developments in the Spectroscopy of Structural Phase Transitions", in Ref. 2, pp. 747-760
18 G.A. Gehring, K.A. Gehring: Co-operative Jahn-Teller effects. Rep. Prog. Phys. *38*, 1-89 (1975)
19 G.A. Gehring: On the observation of critical indices of primary and secondary oder parameters using birefringence. J. Phys. C *10*, 531-542 (1977)
20 I. Hatta, T. Ishikawa: On central mode from a theoretical viewpoint (solid state and ferroelectric transitions). Solid State Phys. *10*, 153-166 (1975) (in Japanese)
21 P.C. Hohenberg, B.I. Halperin: Theory of dynamic critical phenomena. Rev. Mod. Phys. *49*, 435-475 (1977)
22 I.R. Jahn, H. Dachs: Linear birefringence in magnetic phase transitions. Fortschr. Mineral. *51*, 176-239 (1974)
23 F. Jona, G. Shirane: *Ferroelectric Crystals* (Macmillan, New York 1962)
24 S.I. Mizushima: "The Raman Effect", in *Light and Matter II, Encyclopedia of Physics*, Vol. *XXVI*, ed. by S. Flügge (Springer, Berlin, Heidelberg, New York 1971) pp. 171-243
25 K.A. Müller, A. Rigamonti (eds.): *Proceedings of the International School of Physics 'Enrico Fermi', Course LIX. Local Properties at Phase Transitions* (North-Holland, Amsterdam 1976)
26 T. Nakamura: Light scattering studies on soft phonon phase transitions. Ferroelectrics *9*, 159-169 (1975)
27 P.S. Peercy: "Pressure Dependence of Inelastic Light Scattering from Soft Mode Systems", in Ref. 2, pp. 782-795
28 J. Petzelt, J. Fousek: Structural phase transitions in crystals. Czech J. Phys. A *26*, 337-365 (1976)

29 W. Richter: "Resonant Raman Scattering in Semiconductors", in *Solid State Physics*, Springer Tracts in Modern Physics, Vol. 78 (Springer, Berlin, Heidelberg, New York 1976) pp. 121-272
30 T. Riste (ed.): *Anharmonic Lattices, Structural Transitions and Melting* (Noordhoff, Leiden 1974)
31 T. Riste (ed.): *Electron-Phonon Interactions and Phase Transitions* (Plenum, New York 1978)
32 E.J. Samuelson, E. Anderson, J. Feder (eds.): *Structural Phase Transitions and Soft Modes* (Universitetsforlaget, Oslo 1971)
33 J.F. Scott: Soft mode spectroscopy: Experimental studies of structural phase transitions. Rev. Mod. Phys. *46*, 83-128 (1974)
34 J.F. Scott: "Vibrational Raman Spectroscopy as a Probe of Solid State Structure and Structural Phase Transitions", in *Vibrational Spectra and Structure*, Vol. 5 (Elsevier, New York 1976) pp. 67-100
35 G.A. Smolensky, Z.M. Hashkozev: Light scattering by acoustic phonons in ferroelectric and piezoelectric crystals. Mater. Res. Bull. *6(10)*, 1065-1074 (1971)
36 E.F. Steigmeier: Light scattering near structural phase transitions. Ferroelectrics *7*, 65-70 (1974)
37 H. Vogt: Study of structural phase transitions by techniques of nonlinear optics. Appl. Phys. *5*, 85-96 (1974)
38 J.M. Worlock: "Light Scattering Studies of Structural Phase Transitions", in Ref. 32, pp. 329-332
39 G.B. Wright (ed.): *Light Scattering Spectra of Solids* (Springer, Berlin, Heidelberg, New York 1969)

2 *Instrumentation*

40 J.R. Birch, C.E. Bulleid: Phase errors in dispersive fourier transform spectroscopy. Infrared Phys. *17*, 279-282 (1977)
41 D.S. Cannel, G.B. Benedek: Brillouin spectrum of xenon near its critical point. Phys. Rev. Lett. *25*, 1157 (1970)
42 J.J. Chamberlain, E. Gibbs, H.A. Gebbie: Refractometry in the far IR using a two-beam interferometer. Nature *198*, 874 (1963)
43 J.P. Dougherty, S.K. Kurtz: A second harmonic analyzer for the detection of noncentrosymmetry. J.Appl. Crystallogr. *9*, 145-158 (1976)
44 L.A. Firstein, J.M. Cherlow, R.W. Hellwarth: Observation of dynamical central peak in the light-scattering spectrum of a glass. Appl. Phys. Lett. *28*, 25-27 (1976)
45 P.R. Griffiths: *Chemical Infrared Fourier Transform Spectroscopy* (Wiley, New York 1975)
46 R. Hirsch, G. Stuhmer: A simple technique to measure optical SHG in powder samples by means of electronic integration (application to phase transition investigations). Phys. Status Solidi A *37*, 105-108 (1976)
47 S. Itoh, T. Nakamura: Brillouin scattering study of $Gd_2(MoO_4)_3$ using a double Fabry-Perot interferometer. Jpn. J. Appl. Phys. *14*, (Suppl) 183-186 (1975)
48 E.H. Izen, F.A. Modine: A combined magnetic circular dichroism and EPR spectrometer. Rev. Sci. Instrum. *43*, 1563 (1972)
49 A.R. Johnston: A polarimeter for measurement of transient retardation changes. Appl. Opt. *8*, 1837-1848 (1969)
50 M.F. Kimmitt: Recent developments in IR detectors. Infrared Phys. *17*, 459-466 (1977)
51 K.H. Langley, N.C. Ford: Attenuation of the Rayleigh component in Brillouin spectroscopy using interferometric filtering. J. Opt. Soc. Am. *59*, 281-284 (1969)
52 J.B. Lastovka: An optical apparatus for very-small-angle light scattering--design, analysis, and performance. Bell Syst. Tech. J. *55*, 1225 (1976)
53 D.A. Ledsham, W.G. Chambers, T.J. Parker: Dispersive reflection spectroscopy in the far infrared using a polarising interferometer. Infrared Phys. *16*, 515-522 (1976)
54 D.A. Ledsham, W.G. Chambers, T.J. Parker: Far infrared measurements on KDP using dispersive reflection spectroscopy. Infrared Phys. *17*, 165-172 (1977)

55 G. Leveque: Reflectivity extrapolations in Kramers-Kronig analysis. J. Phys. C *10*, 4877 (1977)
56 K.B. Lyons, P.A. Fleury: Digital normalization of iodine filter structure in quasielastic light scattering. J. Appl. Phys. *47*, 4898-4900 (1976)
57 A.J. Mccaffery, P.D. Rowan, R.A. Shatwell: The detection of crystallographic phase transitions by linear dichroism. J. Phys. C *6*, L387-L389 (1973)
58 F. Micheron, G. Bismuth: Real time measurement of induced variation of birefringence in transparent media. Rev. Sci. Instrum. *43*, 292 (1972)
59 F.A. Modine, R.W. Major, E. Sonder: High frequency polarization modulation method for measuring birefringence. Appl. Opt. *14*, 757 (1975)
60 K.F. Pai, T.J. Parker, R.P. Lowndes: "The Temperature Dependence of the Dielectric Response of Pseudo Displacive Ferroelectrics by Dispersive Fourier Transform Spectroscopy", in *2nd Int. Conf. and Winter School on Submillimeter Waves and their Applications*, San Juan, Puerto Rico, 6-11 Dec 1976 (IEEE, 1976) pp. 176 et seq.
61 T.J. Parker, W.G. Chambers: Dispersion reflection spectroscopy in the far IR by division of the field of view in a Michelson interferometer. Infrared Phys. *14*, 207-215 (1974)
62 T.J. Parker, D.A. Ledsham, W.G. Chambers: Study of soft mode behaviour in KDP by dispersive Fourier transform spectroscopy. Ferroelectrics *14*, 625 (1976)
63 T.S. Robinson, W.C. Price: The determination of IR absorption spectra from reflection measurements. Proc. Phys. Soc. B*66*, 969-974 (1953)
64 C. Roychoudhuri, M. Hercher: Stable multipass Fabry-Perot interferometer: Design and analysis, Appl. Opt. *16*, 2514-2520 (1977)
65 J.R. Sandercock: Some recent developments in Brillouin scattering. RCA Rev. *36*, 89-107 (1975)
66 J. Sandercock: Light scattering from surface acoustic phonons. Bull. Am. Phys. Soc. *23*, 387-388 (1978)
67 N. Sheppard, R.G. Greenler, P.R. Griffiths: A theoretical discussion of the comparative performances of dispersion and Fourier transform spectrometers for the IR region. Appl. Spec. *31*, 448-450 (1977)
68 M. Singh: Theory of holographic plane gratings. Indian J. Pure Appl. Phys. *15*, 338 (1977)
69 L.J. Soltzberg: High path-difference optical compensator: Its use in the observation of solid-solid phase transformations. Appl. Opt. *14*, 1664-1666 (1975)
70 P.R. Staal, J.E. Eldridge: Improvements in a dispersion reflection spectrometer using a commercial Michelson interferometer. Infrared Phys. *17*, 299-303 (1977)
71 C.H.F. Velzel: On the imaging properties of holographic gratings. J. Opt. Soc. Am. *67*, 1021-1027 (1977)

3 *Theory*

72 A. Abragam, B. Bleaney: *Electron Paramagnetic Resonance of Transition Ions* (Clarendon, Oxford 1970) Chap. 16
73 A. Aharony, A.D. Bruce: Polycritical points and floplike displacive transitions in perovskites. Phys. Rev. Lett. *33*, 427-430 (1974)
74 K. Aizu: General consideration of ferroelectrics and ferroelastics such that the electric susceptibility or elastic compliance is temperature independent in the prototypic phase. J. Phys. Soc. Jpn. *33*, 629-634 (1972)
75 K. Aizu: Phenomenological Theory of the transition sequence in K_2SeO_4, Pnam: An incommensurate phase: commensurate $Pna2_1$. J. Phys. Soc. Jpn. *43*, 188-194 (1977)
76 K.S. Aleksandrov: The sequences of structural phase transitions in perovskites. Ferroelectrics *14*, 801-805 (1976)
77 S. Aubry, R. Pick: Dynamical behaviour of a coupled double-well system. Ferroelectrics *8*, 471-473 (1974)
78 A.S. Barker: Long-wavelength soft modes, central peaks, and the Lyddane-Sachs-Teller relation. Phys. Rev. B *12*, 4071-4084 (1975)
79 H. Beck: On the dynamics of structural phase transitions. J. Phys. C *9*, 33-49 (1976)

80 K. Binder: Dynamics of clusters near a critical point: central peak and EPR spectra. Solid State Commun. *24*, 401-405 (1977)
81 R. Blinc, V. Smolej, B. Zeks, I. Levstek, B.B. Lavrencic: Dynamical susceptibility of KH_2PO_4 type ferroelectrics below T_c (soft mode spectrum). Ferroelectrics *7*, 203-204 (1974)
82 A.D. Bruce, R.A. Cowley: Renormalization group studies of critical phenomena at structural phase transitions. Ferroelectrics *7*, 397-399 (1974)
83 A.D. Bruce: Structural phase transitions and the renormalization group. Ferroelectrics *12*, 21-29 (1976)
84 A.D. Bruce, R.A. Cowley: "Renormalization Group Studies of Incommensurate Phase Transitions", in *Proc. of Intl. Conf. on Lattice Dynamics, Paris, 1977* (Flammarion Paris 1978)
85 L.N. Bulaevskii: Structural transitions with formation of charge-density waves in layer compounds. Sov. Phys.-Usp. *19*, 836-843 (1976)
86 W. Cochran: Crystal stability and the theory of ferroelectricity. Adv. Phys. *9*, 387 (1960)
87 W. Cochran: Crystal stability and the theory of ferroelectric-piezoelectric crystals. Adv. Phys. *10*, 401 (1961)
88 E. Courtens: Rayleigh scattering and critical dynamics at structural transitions in perovskites. Phys. Rev. Lett. *37*, 1584-1587 (1976)
89 E. Courtens: "The Observation of Birefringence in Fluctuating Anisotropic Media. Application to the 105K Phase Transition in $SrTiO_3$", in Ref. 25, pp. 293-311
90 E. Courtens: Deuterium impurities and the Rayleigh central peak in hydrogen-bonded ferroelectrics. Phys. Rev. Lett. *39*, 561-564 (1977)
91 R.A. Cowley, A.D. Bruce: Application of the Wilson theory of critical phenomena to a structural phase transition. J. Phys. C *6*, L191-L196 (1973)
92 R.A. Cowley, G.J. Coombs: Paraelectric, piezoelectric, and pyroelectric crystals: II. Phase transitions. J. Phys. C *6*, 143 (1973)
93 R.A. Cowley: Acoustic phonon instabilities and structural phase transitions. Phys. Rev. B *13*, 4877-4885 (1976)
94 V. Dvorak: Improper ferroelectrics. Ferroelectrics *7*, 1-9 (1974)
95 R.J. Elliott, R.T. Harley, W. Hayes, S.R.P. Smith: Raman scattering and theoretical studies of Jahn-Teller induced phase transitions in some rare earth compounds. Proc. R. Soc. London A *328*, 217-266 (1972)
96 F. Gervais: Anharmonicity near structural phase transitions. Ferroelectrics *13*, 555-557 (1976)
97 H. Granicher, K.A. Müller: On the nature of phase transitions and nomenclature. Mater. Res. Bull *6*, 977-988 (1971)
98 B.I. Halperin, C.M. Varma: Defects and the central peak near structural phase transitions. Phys. Rev. B*14*, 4030-4044 (1976)
99 W. Hasenfratz, R. Klein, N. Theodorakopoulos: Light scattering from moving cluster walls in a model of displacive phase transitions. Solid State Commun. *18*, 893-895 (1976)
100 J.B. Hastings, S.M. Shapiro, B.C. Frazer: Central peak enhancement in hydrogen reduced $SrTiO_3$. Phys. Rev. Lett. *40*, 237-240 (1978)
101 V. Janovec, V. Dvorak, J. Petzelt: Symmetry classification and properties of equi-translation structural phase transitions. Czech. J. Phys. B*25*, 1362-1396 (1975)
102 A.I. Larkin, D.E. Khmel'nitskii: Phase transitions in uniaxial ferroelectrics. Sov. Phys.-JETP *29*, 1123 (1969)
103 A.P. Levanyuk, D.G. Sannikov: Theory of ferroelectricity in GMO. Sov. Phys.-Solid State *12*, 2418 (1971)
104 A.P. Levanyuk, N.V. Shchedrina: Theory of damping of optical phonons near phase transition points. Sov. Phys.-Solid State *16*, 923-925 (1974)
105 A.P. Levanyuk, V.V. Osipov: Theory of an optically induced change in the refractive index. Sov. Phys.-Solid State *17*, 2340-2342 (1975)
106 A.P. Levanyuk: On scattering of light near phase transition points. Sov. Phys.-JETP *43*, 652 (1976)
107 A.P. Levanyuk, D.G. Sannikov: Theory of phase transitions in ferroelectrics accompanied by the formation of a superstructure whose period is not equal to a multiple of the initial period. Sov. Phys.-Solid State *18*, 245-248 (1976)

108 R. Loudon: The Raman effect in crystals. Adv. Phys. *13*, 423 (1964)
109 M. Matsushita: Anomalous temperature dependence of the frequency and damping constant of phonons near T_λ in ammonium halides. J. Chem. Phys. *65*, 23-28 (1976)
110 W.L. McMillan: Time dependent Landau theory of CDW in transition metal dichalcogenides. Phys. Rev. B*12*, 1187-1199 (1975)
111 W.L. McMillan: Theory of discommensurations and the commensurate-incommensurate CDW phase transition. Phys. Rev. B*14*, 1496-1502 (1976)
112 W.L. McMillan: Microscopic model of charge density waves in 2H-$TaSe_2$. Phys. Rev. B*16*, 643-650 (1977)
113 W.L. McMillan: Collective modes of a charge density wave near the lock-in transition. Phys. Rev. B*16*, 4655-4658 (1977)
114 L. Merten, G. Borstel: Phenomenological theory of coupling of Debye relaxations with IR-active phonons. Phys. Status Solidi B *77*, 221-227 (1976)
115 K.A. Muller, W. Berlinger: Static critical exponents at structural transitions. Phys. Rev. Lett. *26*, 13-16 (1971)
116 K.K. Murata: Mean-field calculation of the central peak due to the phonon-ising dynamic crossover. Phys. Rev. B *11*, 462-477 (1975)
117 R. Opperman: A microscopic calculation of dynamic critical exponents for displacive phase transitions. Z. Phys. B *28*, 92-102 (1977)
118 J. Petersson: A dynamical Landau theory of ferroelectrics with coupled order parameters. Ferroelectrics *13*, 565-567 (1976)
119 J. Petzelt, V. Dvorak: Changes of infrared and Raman spectra induced by structural phase transitions. I. General considerations. J. Phys. C *9*, 1571-1586 (1976)
120 J. Petzelt, V. Dvorak: Changes of infrared and Raman spectra induced by structural phase transitions. II. Examples. J. Phys. C *9*, 1587-1601 (1976)
121 E. Pytte, J. Fedder: Theory of a structural phase transition in perovskite type crystals. Phys. Rev. *187*, 1077 (1969)
122 E. Pytte: Model for ferroelectric GMO. Solid State Commun. *8*, 2101-2104 (1970)
123 E. Pytte: Dynamics of Jahn-Teller phase transitions. Ferroelectrics *7*, 193 (1974)
124 E. Pytte: "Acoustic Anomalies at Structural Phase Transitions", in Ref. 32, pp. 171-188
125 G. Reiter, N. Tzoar: Critical fluctuations in the weakly anharmonic crystal model of a structural phase transition. Phys. Rev. B *14*, 208-218 (1976)
126 J. Ruvalds, E.N. Economou, K.L. Ngai: "Phonon Coupling and the Central Peak Anomaly in Condensed Matter", in Ref. 2, pp. 770-774
127 H. Schmidt, D. Schwabl: Localized modes and central peak at displacive phase transitions. Phys. Lett. A *61a*, 476-478 (1977)
128 T. Schneider, E. Stoll: New excitations in systems undergoing displacive antiferrodistortive structural phase transitions. J. Phys. C *8*, 283-288 (1975)
129 T. Schneider, E. Stoll: Molecular-dynamics study of structural-phase transitions. I. One-component displacement models. Phys. Rev. B *13*, 1216-1237 (1976)
130 T. Schneider. E. Stoll: "Molecular-Dynamics Investigation of Structural Phase Transitions", in Ref. 25, pp. 344-368
131 F. Schwabl: Critical dynamics and microscopic theory of the central peak at structural phase transitions. Ferroelectrics *7*, 395-396 (1974)
132 T. Shigenari: Soft modes and their symmetries. Solid State Phys. *10*, 381-390 (1975) (in Japanese)
133 N. Szabo: On the coupling between order parameter and temperature fluctuations in structural phase transitions. J. Phys. C *9*, 259-267 (1976)
134 H. Thomas, K.A. Muller: Structural phase transitions in perovskite type crystals. Phys. Rev. Lett. *21*, 1256 (1968)
135 H. Thomas: "Mean-Field Theory of Phase Transitions", in Ref. 25, pp. 3-44
136 H. Thomas: "Discussion Contribution to the Critical Dynamics at Phase Transitions", in Ref. 25, pp. 369-373
137 J.C. Toledano: "Ferroelasticite", Ann. Telecommun. *29*, 1 (1974)
138 P. Toledano, J.C. Toledano: Order-parameter symmetries for improper ferroelectric nonferroelastic transitions. Phys. Rev. B *14*, 3097-3109 (1976)
139 C.M. Varma: Dynamics of anharmonic lattices: Solitons and the central-peak problem in one dimension. Phys. Rev. B *14*, 244-254 (1976)

140 K.S. Alexandrov, A.T. Anistratov, D.X. Blat, L.I. Zherebtsova, V.I. Zinenko, I.M. Iskornev, S.V. Melnikova, I.N. Flerov: Properties of NH_4HSO_4 and $RbHSO_4$ single crystals near their Curie points. Ferroelectrics *12*, 191-193 (1976)
141 P. Bastie, J. Bornarel, J. Lajzerowicz, J.F. Legrand: Induced and spontaneous strain in ferroelectric ferroelastic tanane. Ferroelectrics *13*, 319-321 (1976)
142 P. Bastie. J. Bornarel, J.F. Legrand: Perturbated regions in the vicinity of domain wall in ferroelastic-ferroelectric crystals. Ferroelectrics *13*, 455-458 (1976)
143 D.J. Benard, W.C. Walker: Birefringence of $PrAlO_3$ in the vicinity of 118K. Solid State Commun. *18*, 717-719 (1976)
144 J.D. Bierlein, A.W. Sleight: Ferroelasticity in $BiVO_4$. Solid State Commun. *16*, 69-70 (1975)
145 L.H. Brixner, R.B. Flippen, W. Jeitschko: Preparation and properties of the $Pb_3(PO_4)_2$ - $Pb_3(VO_4)_2$ system. Mater.Res.Bull. *10*, 1327-1334 (1975)
146 E. F. Dudnik, E.V. Sinyakov, V.V. Gene, I.E. Mnushkina: Spontaneous elastooptic effect in lead orthophosphate. Sov. Phys.-Solid State *17*, 793 (1975)
147 E.F. Dudnik, E.V. Sinyakov, V.V. Gene, S.V. Vagin: Domain structure in ferroelastic lead orthophosphate. Sov. Phys.-Solid State *17*, 1212-1213 (1975)
148 K.H. Germann, W. Muller-Lierheim, H.H. Otto, T. Suski: High temperature phase transition in $Pb_5Ge_3O_{11}$. Phys. Status Solidi A *35*, K165-K167 (1976)
149 M. Glogarova, J. Fousek: Dielectric, optical, and domain properties of the improper ferroelectric $(NH_4)_2Cd_2(SO_4)_3$. Phys. Status Solidi A *15*, 579-590 (1973)
150 B.N. Grib, I.I. Kondilenko, P.A. Korotkov, A.I. Pisanskii, Yu.P. Tsyshchenko: Nature of the electro-optic effect and vibration spectrum of a KH_2PO_4 crystal. Opt. Spec. *38*, 415-418 (1975)
151 R.T. Harley, W. Hayes, A.M. Perry, S.R.P. Smith: Jahn-Teller phase transitions in $Tb_cGd_{1-c}VO_4$. Solid State Commun. *14*, 521-524 (1974)
152 R.T. Harley, R.M. Macfarlane: A determination of the critical exponent β in $TbVO_4$ and $DyVO_4$ using linear birefringence. J. Phys. C *8*, L451-L455 (1975)
153 S. Hirotsu: Experimental studies of structural phase transitions in $CsPbCl_3$ J. Phys. Soc. Jpn. *31*, 552-560 (1971)
154 U.T. Hoechli, D.W. Pohl: Birefringence in the surface layer of cubic $BaTiO_3$. Ferroelectrics *13*, 403-405 (1976)
155 U.T. Hoechli: "Critical Dielectric and Optical Properties and Storage in Incipient Ferroelectrics at 4K", in *Proc. of the 6th Int. Cryogenic Engineering Conf.*, Grenoble, France, 11-14 May, 1976, ed. by F.R.S. Mendelssohn (IPC Sci. and Technology Press, 1976) pp. 333-336
156 Z. Iqbal, C.W. Christoe: Raman scattering study of phase transition in ammonium sulfate. Ferroelectrics *12*, 177-179 (1976)
157 I.R. Jahn, I.H. Brunskill, H. Dacts, R. Bausch: The β,γ phase transition in the system $NH_4Cl_{1-x}Br_x$. J. Phys. C *8*, 3280-3286 (1975)
158 I.R. Jahn, I.H. Brunskill: Nature of the β,γ phase transition in $NH_4Cl_{1-x}Br_x$ mixed crystals. Acta Crystallogr. A *A31*, S189 (1975)
159 A.R. Johnston, J.M. Weingart: Determination of the low-frequency linear electro-optic effect in tetragonal $BaTiO_3$. J. Opt. Soc. Am. *55*, 828 (1965)
160 J. Kobayashi: Peculiar physical properties of improper ferroelectrics. Ferroelectrics *10*, 277-282 (1976)
161 D.S.H. Kovach, A.N. Borets, I.D. Turyan: Interference investigations into the phase transitions in SbSI. Ukr. Fiz. Zh. *19*, 2058-2061 (1974) (in Russian)
162 N.N. Krainik, V.A. Trepakov, L.S. Kamzina, D.G. Sakharov, B.A. Volik, V.A. Pis'mennyi, K.P. Skornyakova: Characteristic features of the electrooptic effect in lead magnoniobate subjected to strong alternating electric fields in the region of a broad phase transition. Sov. Phys.-Solid State *17*, 122-124 (1975)
163 N.N. Krainik, G.A. Smolensky, L.S. Kamzina, V.A. Trepakov, A.A. Berezhnoi: The influence of polarization orientation processes on the optical and electrooptic properties of ferroelectrics with a diffuse phase transition. Ferroelectrics with a diffuse phase transition. Ferroelectrics *12*, 169-171 (1976)
164 L.A. Pozdnyakova: Twinning in $CsPbCl_3$ crystals (and birefringence). Sov. Phys. -Crystallogr. *19*, 552 (1975)

165 P. Seidel, W. Hoffmann: X-ray diffraction studies, optical and dielectrical measurements of sodium niobate in the range of temperature from 293K to 10K. Ferroelectrics *12*, 203 (1976)
166 G.A. Smolensky, N.N. Krainik, L.S. Kamzina, A.A. Karamyan, I.E. Mylnikova: Electro-optical and some optical properties of lead magnoniobate crystals. Ferroelectrics *7*, 99-100 (1974)
167 L.J. Soltzberg, P.A. Piliero, M.R. Shea: Optical study of the phase transition in phenanthrene single crystals. Mol. Cryst. Liq. Cryst. *29*, 151-154 (1974)
168 J.M. Thierry, E. Coquet, J.M. Crettez: Interferometric measurements of thermal expansion coefficients and birefringence of a α-$LiIO_3$ with temperature. Opt. Commun. *16*, 417-419 (1976)
169 J.C. Toledano, J. Schneck: Diffused ferroelastic phase transition in barium sodium niobate. Solid State Commun. *16*, 1101-1104 (1975)
170 J. Torres, J. Aubree, J. Brandon: Refractive indices and spontaneous birefringence of ferroelastic lead orthophosphate. Opt. Commun. *12*, 416-417 (1974);
– M. Vallade: Simultaneous measurements of the second harmonic generation and the birefringence of KH_2PO_4 near its ferroelectric point. Phys. Rev. B *12*, 3755-3765 (1975)
171 R. Verreault: Crystallographic, optical and magnetic properties of Eu_2SiO_4. Phys. Kondens. Mater. *14*, 37-54 (1971);
– K. Viswanathan, E. Salje: Thermal contraction in NH_4IO_3 and its significance with respect to the phase transition. Acta Crystallogr. A *A31*, 810-813 (1975)
172 M.P. Zaitseva, A.T. Anistratov, A.E. Krumin, L.A. Shabanova, I.M. Iskornev: Ferroelectric transition in deuterated aluminium methyl ammonium alums. Sov. Phys.-Crystallogr. *19*, 632-634 (1975)

5 *Zone Center Transitions: Optical Absorption and Fluorescence*

173 E.R. Bernstein, K.M. Chen: Spectroscopic properties of rare earth borohydrides: $Er(BH_4)_3 \cdot 3ThF$ in pure and mixed crystals. Chem. Phys. *10*, 215-228 (1975)
174 E.V. Bursian, Ya. G. Girshberg, A.V. Ruzhnikov: The correlation between optical absorption spectra, carrier mobility, and phase transition temperature in some ferroelectrics. Phys. Status Solidi B *74*, 689-693 (1976)
175 T.J. Glynn, R.T. Harley, W. Hayes, A.J. Rushworth, S.H. Smith: The phase transitions in $Pr_pLa_{1-p}AlO_3$. J. Phys. C *8*, L126-L128 (1975);
B.N. Grib, I.I. Kondilenko, P.A. Korotkov, A.I. Pisanskii, Yu. P. Tsyashchenko: Nature of the electro-optic effect and vibration spectrum of a KH_2PO_4 crystal. Opt. Spec. *38*, 415-418 (1975)
176 R.T. Harley, W. Hayes, A.M. Perry, S.R.P. Smith: The phase transitions of $PrAlO_3$. J. Phys. C *6*, 2382-2400 (1973)
177 A.A. Kaminskii, V.A. Koptsik, Yu.A. Markaev, I.I. Naumova, L.N. Rashkovich, S.E. Sarkisov: Stimulated emission of Nd^{3+} ions in a ferroelectric barium-sodium niobate crystal. Sov. Tech. Phys. Lett. *1*, 205-206 (1975)
178 V.I. Lazorenko, B.M. Rud, Yu. B. Paderno, L.A. Dvorina: Some physical properties of yttrium silicides. Inorg. Mater. *10*, 1150-1151 (1974)
179 K. Mamola, R. Wu: Effects of pressure of optical F-bands in rubidium halides. J. Phys. Chem. Solids *37*, 683-688 (1976)
180 S.K. Manlief, H.Y. Fan: Absorption edge splitting in $KTa_xNb_{1-x}O_3$ (Paraelectric and ferroelectric phases). Phys. Rev. B *6*, 185-192 (1972)
181 K. Miyauchi, G. Toda: Effects of crystal-lattice distortion on optical transmittance of the $(Pb,La)(Zr,Ti)O_3$ system. J. Am. Ceram. Soc.*58*, 361-363 (1975)
182 K. Nako, M. Balkanski: Electronic band structures of SbSI in the para- and ferroelectric phases. Phys. Rev. B *8*, 5759-5780 (1973)
183 V.K. Novik, N.F. Karyakina, V.A. Timoshenkov: Intrinsic absorption and thermal properties of stibiotantalite. Sov. Phys.-Solid State *14*, 1288-1290 (1972)
184 V.V. Panfilov, S.I. Subbotin, L.F. Vereshchagin, I.I. Ivanov, R.T. Molchanova: On exciton absorption, band structure, and phase transformation of GaSe under pressure. Phys. Status Solidi B *72*, 823-831 (1975)
185 N. Presser, M. Ratner, R. Sundheim: Dependence of optical lifetime on thermal transitions in some tetraalkylammonium tetrahalometalates. Chem. Phys. Lett. *45*, 572-574 (1977)

186 V.G. Savitskii, R.N. Kovtun, V.E. Aleksyuk, P.F. Protsakh: Temperature dependence of the forbidden-band width of $Ba_{0.5}Sr_{0.5}Nb_2O_6$ single crystals in the phase transition region. Sov. Phys.-Solid State *15*, 2464 (1974)
187 R.W. Schwartz, S.F. Watkins, C.J. O'Connor, R.L. Carlin: Low temperature crystalline phase transition in some elpasolite-hexachlorides. (optical and EPR spectra). J. Chem. Soc. Faraday Trans. II *72*, 565-570 (1976)
188 M.D. Sturge, E. Cohen, L.G. van Uitert, R.P. van Stapele: Microscopic order parameters in $PrAlO_3$. Phys. Rev. B *11*, 4768-4779 (1975)
189 Z.S. Vasilina, V.M. Varikash, I.S. Zheludev, N.A. Romanyuk: Temperature-induced changes in the fundamental reflection spectra of crystals isomorphous with triglycine sulphate. Sov. Phys.-Solid State *15*, 1665 (1974);
– R. Verreault: Crystallographic, optical and magnetic properties of Eu_2SiO_4. Phys. Kondens. Mater. *14*, 37-54 (1971)
190 M.D. Volnyanskii, A.Yu. Kudzin, Yu.I. Samchenko, S.A. Flerova: Luminescence of $Pb_5Ge_3O_{11}$ crystals under pulse polarisation switching conditions. Sov. Phys.-Solid State *18*, 339 (1976)
191 P.J. Wang, H.G. Drickamer: Transformation from tetrahedral to planar symmetry in Cs_2CuCl_4 and Cs_2CuBr_4 at high pressure. J. Chem. Phys. *59*, 559-560 (1973)
192 E. Wiesendanger: Dielectric, mechanical and optical properties of orthorhombic $KNbO_3$. Ferroelectrics *6*, 263-281 (1974)
193 M.A. Yakubovskii, L.M. Rabkin, D.S. Kohevskaya, E.G. Fesenko: Optical properties of lead titanate single crystals in the region of the intrinsic absorption edge. Sov. Phys.-Crystallogr. *19*, 541-542 (1975)
194 A.Kh. Zeinally, A.A. Agasiev, Sh.M. Efendiev: Direct 'allowed' transitions in SbSI. Sov. Phys.-Semicond. *8*, 128-129 (1974)
195 A.Kh. Zeinally, N.N. Lebedeva, B.S. Agaronov, A.M. Safarov, M.K. Sheinkman: Characteristics of the photoconductivity of barium-strontium niobate in the phase transition region. Sov. Phys.-Solid State *18*, 363-364 (1976)

6 *Zone Center Transitions: IR Absorption and Light Scattering*

196 P.K. Acharya, P.S. Narayanan: Vibration spectra of ferroelectric sulphates. I. Indian J. Pure Appl. Phys. *11*, 514-518 (1973)
197 P.K. Acharya, P.S. Narayanan: Vibration spectra of ferroelectric sulphates. II. $RbHSO_4$, NH_4HSO_4 and $(NH_4)_3H(SO_4)_2$. Indian J. Pure Appl. Phys. *11*, 519-521 (1973)
198 D.K. Agrawal, C.H. Perry: Long-wavelength optical phonons and phase transitions in SbSI. Phys. Rev. B *4*, 1893-1902 (1971)
199 N. Amer, P.Y. Yu, Y. Petroff, Y.R. Shen: "Resonant Raman Scattering in SbSI Near the Absorption Edge", in Ref. 2, pp. 49-53
200 A.T. Anistratov, S.V. Mel'nikova, S.V. Komogortsev: Electro-optical and thermo-optical properties of ferroelectric $Ca_2Sr(C_2H_5CO_2)_6$ crystals. Sov. Phys.-Crystallogr. *20*, 522-523 (1975)
201 H. Arend, R. Blinc, A. Kandusar: Search for ferroelectricity in the $PbHPO_4$ family. Ferroelectrics *13*, 511-513 (1976)
202 J. Azoulay, D. Gerlich, E. Wiener, I. Pelah: Brillouin scattering studies of KH_2PO_4, CsH_2AsO_4 and the deuterated ferroelectric crystals. Ferroelectrics *8*, 599-602 (1974)
203 J. Azoulay, D. Gerlich, E. Wiener-Avnear, I. Pelah: Piezoelectric coupled soft mode in CsH_2AsO_2 and CsD_2AsO_4 studied by Brillouin scattering. Phys. Status Solidi B *81*, 295-306 (1977)
204 D.F. Baisa, A.I. Barabash, A.T. Belokon', G.A. Puchkovskaya, G.K. Semin, Yu.A. Frolkov: Investigation into the dynamics of the phase transitions in a KIO_3 crystal using NQR and IR spectroscopy methods. Izv. Akad. Nauk SSSR Ser. Fiz. *39*, 702-707 (1975)
205 A. Bartzokas, D. Siapkas: Optical phonons and phase transitions of SbSSr-SbSI mixed crystals. Ferroelectrics *12*, 127-129 (1976)
206 D. Bauerle, L. Genzel, T.P. Martin: Ferroelectric phase transition in small crystals of $BaTiO_3$. Phys. Status Solidi B *59*, 459-464 (1973)
207 D. Bauerle, R. Clarke, T.P. Martin: The ferroelectric-paraelectric phase transition in $PbTi_{0.1}Zr_{0.9}O_3$. Phys. Status Solidi B *71*, K173-K176 (1975)

208 W. Bauhofer, L. Genzel, C.H. Perry, I.R. Jahn: "Optical Phonons and Phase Transitions in $(NH_4)Cl_{1-x}Br_x$ Mixed Crystals", in Ref. 2, pp. 918-922
209 M.V. Belousov, V.A. Kamyshev, A.A. Shuttin: Vibrational spectrum and nature of the phase transition in an ammonium sulphate crystal. Bull. Acad. Sci. USSR: Phys. Ser. *39*, (4): 90 (1975)
210 J.P. Benoit: Raman effect of α and β $Pb_3(PO_4)_2$. Ferroelectrics *13*, 331-332 (1976)
211 R. Blinc, H. Arend, A. Kanduser: Vibrational spectra, hydrogen bonding, and ferroelectricity in the $PbHPO_4$ family. Phys. Status Solidi B *74*, 425-435 (1976)
212 R. Blinc, R. Pirc, B. Zeks: Soft-mode dynamics of partially deuterated KH_2PO_4-type crystals. Phys. Rev. B *13*, 2943-2949 (1976)
213 A.M. Bon, C. Benoit, J. Giordano: Dynamical properties of crystals of $Sr(NO_3)_2$, $Ba(NO_3)_2$, and $Pb(NO_3)_2$ II, temperature dependence of the infrared spectra", Phys. Status Solidi B *78*, 453-464 (1976)
214 D.G. Bozinis, J.P. Hurrell: Optical modes and dielectric properties of ferroelectric orthorhombic $KNbO_3$. Phys. Rev. B *13*, 3109-3120 (1976)
215 D.G. Bozinis, A. Scalabrin: "Temperature Dependent Raman Scattering and Dielectric Properties of Orthorhombic $KNbO_3$", in Ref. 2, pp. 856-860
216 E.M. Brody, H.Z. Cummins: Brillouin scattering study of the elastic anomaly in ferroelectric KH_2PO_4. Phys. Rev. B*9*, 179-196 (1974)
217 H. Buhay, C.H. Perry: Resonant Raman scattering in the ferroelectric semiconductor SbSi. J. Phys. Chem. *80*, 1208-1211 (1976)
218 G. Burns: Lattice modes in ferroelectric perovskites. II. $Pb_{1-x}Ba_xTiO_3$ including $BaTiO_3$. Phys. Rev. B *10*, 1951-1959 (1974)
219 G. Burns, F.H. Dacol: Coupled mode behaviour of the soft mode in single crystals of several perovskite ferroelectrics. Solid State Commun. *18*, 1325-1328 (1976)
220 G. Burns: Consideration of a frequency-independent damping coefficient from Raman measurements. Phys. Rev. Lett. *37*, 229-232 (1976)
221 E.V. Bursian, J.A.G. Girshberg, A.V. Ruzhnikov: The correlation of parameters characterizing the phase transition and electronic processes in some ferroelectrics. Ferroelectrics *13*, 371 (1976)
222 D.S. Bystrov, A.P. Vorob'ev, E.A. Popova: Nonclassical modes in KH_2PO_4 and KD_2PO_4 crystals. Bull. Acad. Sci. USSR: Phys. Ser. *39*, (4): 84 (1975)
223 A.N. Cao-Xuan, G. Hauret, J.-P. Chapelle: Brillouin spectra for β phase $Pb_3(PO_4)_2$. C.R. Acad. Sci. B *280*, 543-546 (1975) (in French)
224 A.N. Cao-Xuan, J.P. Chapelle: Brillouin scattering in $Pb_3(PO_4)_2$ (ferroelastic transition). Ferroelectrics *13*, 329 (1976)
225 C. Caville: A low temperature phase transition study of orthorhombic hydrazonium sulphate by Raman spectroscopy. Solid State Commun. *21*, 475-477 (1977)
226 F. Cerdeira, W.B. Holzapfel, D. Bauerle: Effect of pressure on the zone-center phonons of $PbTiO_3$ and on the ferroelectric-paraelectric phase transition. Phys. Rev. B *11*, 1188-1192 (1975)
227 P.O. Cervenka: "Light Scattering Studies in Ferroelectric Crystals", Univ. Southern California, Los Angeles, USA, Ph.D. Thesis, Univ. Microfilms, Ann Arbor, Mich., USA. Order No. 74-21463 (1974) 120p
228 P.O. Cervenka, R.A.O. Prasad: The ferroelectric phase transition and dielectric anomaly in tri-glycine sulfate", Ferroelectrics *11*, 511-517 (1976)
229 T.S. Chang: "Nonlinear Optical and Lattice Dynamical Studies of Some Ferroelectric Crystals", Stanford Univ., Calif, USA, Ph.D. Thesis, Univ. Microfilms, Ann Arbor, Mich., USA. Order No. 72-16704 (1972) 168p
230 H. Chihara, N. Nakamura, A. Inaba: Raman and infrared studies of the ferroelectric transition in $NH_4H(ClCH_2CoO)_2$ and $ND_4D(ClCH_2CoO)_2$. J. Phys. Soc. Jpn. *36*, 1710 (1974)
231 C.A. Cody, R.K. Khanna: A new interpretation of the O-H stretching region of some hydrogen bonded ferroelectrics. Ferroelectrics *9*, 251-253 (1975)
232 S.P. Cramer, B. Hudson, D.M. Burland: A polarised single crystal Raman study of the librational phonons in p-diiodobenzene. J. Chem. Phys. *64*, 1140-1145 (1976)
233 G. Dolino, J.P. Bachheimer: Optical study of the α-β transition of quartz. Ferroelectrics *12*, 173-175 (1976)
234 L.N. Durvasula, R.W. Gammon: "Brillouin Scattering in Potassium Dihydrogen Arsenate Near T_C", in Ref. 2, pp. 775-781

235 L.N. Durvasula, R.W. Gammon: Nature of the central-peak light scattering in potassium dihydrogen phosphate. Phys. Rev. Lett. *38*, 1081-1084 (1977);
– R.J. Elliott, R.T. Harley, W. Hayes, S.R.P. Smith: Raman scattering and theoretical studies of Jahn-Teller induced phase transitions in some rare earth compounds. Proc. R. Soc. London A *328*, 217-266 (1972)
236 J.R. Ferraro, U. Fink: Near infrared reflectance spectra and analysis of H_2S frost as a function of temperature. J. Chem. Phys. *67*, 409-413 (1977)
237 P.A. Fleury, J.M. Worlock: "Electric-field-induced Raman scattering in $SrTiO_3$ and $KTaO_3$. Phys. Rev. *174*, 613-623 (1968)
238 P.A. Fleury, P.D. Lazay: Acoustic soft-optic mode interactions in ferroelectric $BaTiO_3$. Phys. Rev. Lett. *26*, 1331 (1971)
239 P.A. Fleury, P.D. Lazay, L.D. van Uitert: Brillouin-scattering evidence for a new phase transition in perovskite crystals: $PrAlO_3$. Phys. Rev. Lett. *33*, 492-495 (1974)
240 P.A Fleury: "Critical Opalescence at Structural Phase Transitions", in *Theory of Light Scattering in Condensed Matter*, ed. by B. Bendow, J.L. Birman, V.M. Agranovich (Plenum, New York 1976)
241 P.A. Fleury, K.B. Lyons: Central-peak dynamics at the ferroelectric transition in lead germanate. Phys. Rev. Lett. *37*, 1088-1091 (1976)
242 I.J. Fritz, H.Z. Cummins: Rayleigh scattering studies of NH_4Cl at high pressures. Phys. Rev. Lett. *28*, 96-99 (1972)
243 B.N. Ganguly, M. Nicol, W.D. Ellenson: Raman spectroscopic study of the high pressure phase transformation of $NaClO_3$. Phys. Status Solidi B*83*, K115-K121 (1977)
244 B.N. Ganguly, M. Nicol: Effect of hydrostatic pressure on the vibrational properties and the structure of $SrWO_4$ and $PbMoO_4$. Phys. Status Solidi B *79*, 617-622 (1977)
– K.H. Germann, W. Muller-Lierheim, H.H. Otto, T. Suski: High temperature phase transition in $Pb_5Ge_3O_{11}$. Phys. Status Solidi A *35*, K165-K167 (1976)
245 F. Gervais, B. Piriou: Temperature dependence of transverse and longitudinal optic modes in the α and β phases of quartz. Phys. Rev. B *11*, 3944-3950 (1975);
– T.J. Glynn, R.T. Harley, W. Hayes, A.J. Rushworth, S.H. Smith: The phase transitions in $Pr_pLa_{1-p}AlO_3$. J. Phy. C *8*, L126-L128 (1975);
– B.N. Grib, I.I. Kondilenko, P.A. Korotkov, A.I. Pisanskii, Yu.P. Tsyashchenko: Nature of the electro-optic effect and vibration spectrum of a KH_2PO_4 crystal. Opt. Spec. *38*, 415-418 (1975)
246 S.D. Hamann: Infrared spectra and phase transitions of solids under pressure. II. High-Temp. - High Pressures *7*, 177-186 (1975);
– R.T. Harley, W. Hayes, A.M. Perry, S.R.P. Smith: The phase transitions of $PrAlO_3$. J. Phys. C *6*, 2382-2400 (1973)
247 R.T. Harley, W. Hayes, A.J. Rushworth, J.F. Ryan: Light scattering study of the onset of disorder in solid electrolytes at high temperatures. J. Phys. C *8*, L530-L534 (1975)
248 R.T. Harley, W. Hayes, A.M. Perry, S.R.P. Smith: Raman study of excitations in $PrAlO_3$. J. Phys. C *8*, L123-L125 (1975)
249 R.R. Hayes, L. Genzel, T.P. Martin, C.H. Perry: Temperature dependence of infrared absorption and Raman scattering by small NH_4Cl crystals. Phys. Status Solidi B *60*, K27-K29 (1973)
250 D. Heiman, S. Ushioda, J.P. Remeika: Frequency dependence of the damping function of the soft E-symmetry phonon in $PbTiO_3$. Phys. Rev. Lett. *34*, 886-889 (1975)
251 T. Hikita, T. Ikeda: Brillouin scattering in ferroelectric NH_4HSO_4. J. Phys. Soc. Jpn. *42*, 351-352 (1977)
252 J.C. Hill, P.V. Mohan: Far-infrared spectra and ferroelectric phase transition of KNO_3. Ferroelectrics *2*, 201-207 (1971);
– S. Hirotsu: Expermental studies of structural phase transitions in $CsPbCl_3$. J. Phys. Soc. Jpn. *31*, 552-560 (1971)
253 K. Hisano, J.F. Ryan: Quasi-elastic scattering in lead germanate ($5PbO \cdot 3GeO_2$). Solid State Commun. *11*, 1745-1749 (1972)
254 K. Hisano, K. Toda: Underdamped soft mode in bismuth titanate ($Bi_4Ti_3O_{12}$). Solid State Commun. *18*, 585-587 (1976)

255 H.D. Hochheimer, E. Spanner, D. Strauch: Phase diagram of ammonium iodide obtained by Raman spectroscopy under hydrostatic pressure. J. Chem. Phys. *64*, 1583-1585 (1976)
256 J. Holvast, J.H.M. Stoelinga, P. Wyder: Far-infrared investigation of the structural phase transitions of some two-dimensional layer compounds. Ferroelectrics *13*, 543-544 (1976)
257 T.-H. Huang, J.C. Decius, J.W. Nibler: Raman and IR spectra of crystalline phosphine in the γ phase. J. Phys. Chem. Solids *38*, 897-904 (1977)
258 P.V. Huong, M. Couzi: "A Raman Spectroscopic Study of Thallium Dihydrogen Phosphates TlH_2PO_4 and TlD_2PO_4", in Ref. 2, pp. 845-849
259 I.A. Iakovlev, T.S. Velichkina, L.F. Mikheeva: Opalescence in the phase transition of quartz. Sov. Phys.-Dokl. *1*, 215 (1956)
260 Z. Iqbal, C.W. Christoe, D.K. Dawson: Infrared absorption and reflection studies of organic radical salt' K $TCNQ^-$. J. Chem. Phys. *63*, 4485-4489 (1975)
261 Z. Iqbal, C.W. Christoe: Raman scattering study of ferroelectric phase transition in ammonium sulphate. Solid State Commun. *18*, 269-273 (1976)
262 Z. Iqbal: Raman scattering and the electronically induced phase transition in $K_3Fe^*(CN)_6$ at 130 K. J. Phys. C *10*, 3533-3543 (1977)
263 W. Kaczmarek, A. Graja: Lattice dynamics study of the solid solution $(Bi_{1-x}La_x)$ FeO_3 by IR spectroscopy. Solid State Commun. *17*, 851-853 (1975)
264 I.P. Kaminow, T.C. Damen: Temperature dependence of the ferroelectric mode in KH_2PO_4. Phys. Rev. Lett. *20*, 1105 (1968)
265 A.A. Karamyan, N.N. Krainik: Vibrational spectrum of a $PbMg_{0.33}Nb_{0.67}O_3$ crystal. Sov. Phys.-Solid State *15*, 1687-1688 (1974)
266 R.S. Katiyar, J.F. Ryan, J.F. Scott: Proton-phonon coupling in CsH_2asO_4 and KH_2asO_4. Phys. Rev. B*4*, 2635 (1971)
267 T. Kawamura, A. Mitsuishi, N. Furuya, O. Shimomura: Optical properties of KD_2PO_4 crystal in the low-wavenumber region (optical phonon spectra). Tech. Rep. Osaka Univ. *24*, 429-441 (1974)
268 E.T. Keve, A.D. Annis: Studies of phases, phase transitions and properties of some PLZT ceramics. Ferroelectrics *5*, 77-89 (1973)
269 K.E. Khaller, E.M. Yarvekyul'g, L.A. Rebane: Raman spectra and phase transitions in lithium iodate. Sov. Phys.-Solid State *18*, 814-816 (1976)
270 A.S. Knyazev, Yu.M. Poplavko, V.P. Zakharov: Vibrational and dielectric spectra of cadmium titanate. Sov. Phys.-Solid State *16*, 1446-1448 (1975)
271 R.S. Krishnan, V.N. Sankaranarayanan: Vibrational spectra and ferroelectric transition of sodium ammonium selenate dihydrate. Ferroelectrics *7*, 287-289 (1974)
272 N. Lagakos, H.Z. Cummins: Preliminary observation of a central peak in the light-scattering spectrum of KH_2PO_4. Phys. Rev. B *10*, 1063-1069 (1974)
273 N. Lagakos, H.Z. Cummins: Experimental test of the relaxing self-energy model in CsH_2AsO_4. Phys. Rev. Lett. *34*, 883-886 (1975)
274 I. Laulicht, R. Ofek: Anomalous isotope effect on Raman linewidths in KH_2AsO_4 and KD_2AsO_4 crystals. J. Raman Spec. *4*, 41-51 (1975)
275 I. Laulicht, N. Luknar: Internal-mode line-broadening by proton jumps in KH_2PO_4. Chem. Phys. Lett. *47*, 237-240 (1977)
276 B.B. Lavrencic, J. Petzelt: Raman study of the ferroelectric phase transition in $PbHPO_4$ and $PbHAsO_4$. J. Chem. Phys. *67*, 3890-3896 (1977)
277 P.D. Lazay, J.G. Lunacek, N.A. Clark, G.B. Benedek: "The Rayleigh Brillouin Spectrum of NH_4Cl", in *Light Scattering Spectra of Solids*, ed. by G.B. Wright (Springer, Berlin, Heidelberg, New York 1969)
278 N. Lehner, K. Strobel, R. Geick, G. Heger: Lattice dynamics in perovskite-type layer structures. FIR studies of $(CH_3NH_3)_2MnCl_4$ and $(CH_3NH_3)_2FeCl_4$. J. Phys. C *8*, 4096-4106 (1975)
279 R.C. Leung, N.E. Tornberg, R.P. Lowndes: E-mode splitting in the ferroelectric phase of hydrogen-bonded ferroelectrics. J. Phys. C *9*, 4477-4490 (1976)
280 R.C. Leung, W.B. Spillman, N.E. Tornberg, R.P. Lowndes: "E Mode Splitting and Pressure Dependence of the Ferroelectric Mode in the Ferroelectric Arsenates", in Ref. 2, pp. 796-800
281 R.C. Leung, N.E. Tornberg, R.P. Lowndes: The pressure dependence of the ferroelectric mode and molecular vibrations of the ferroelectric arsenates. J. Phys. C *10*, 4855 (1977)

282 D.J. Lockwood, J.W. Arthur, W. Taylor, T.J. Hosea: Observation of a central peak in lead germanate by light scattering. Solid State Commun. *20*, 703-707 (1976)
283 D.J. Lockwood: Observation of soft modes in the Raman spectrum of ferroelectric $Cr_3B_7O_{13}Cl$. Solid State Commun. *18*, 115-117 (1976)
284 W.F. Love, H.D. Hochheimer, M.W. Anderson, R.N. Work, C.T. Walker: Brillouin and ultrasonic study of the temperature dependence of the elastic constants in NaCN. Solid State Commun. *23*, 365-368 (1977)
285 R.P. Lowndes, N.C. Tornberg, R.C. Leung: Ferroelectric mode and molecular structure in the hydrogen-bonded ferroelectric arsenates and their isomorphs. Phys. Rev. B*10*, 911 (1974)
286 K.B. Lyons, R.J. Birgeneau, E.I. Blount, L.G. van Uitert: Electronic excitations in $PrAlO_3$. Phys. Rev. B *11*, 891-900 (1975)
287 K.B. Lyons, P.A. Fleury: Dynamic central peaks in a crystalline solid: $KTaO_3$. Phys. Rev. Lett. *37*, 161-164 (1976)
288 K.B. Lyons, P.A. Fleury: Light scattering investigations of the ferroelectric phase transition in $Pb_5Ge_3O_{11}$. Phys. Rev. B*17*, 2403 (1978)
289 R. Merlin, J.A. Sanjurjo, A. Pinczuk: Coupled mode behavior of the soft E(TO) phonon of $Pb(Ti_{1-x},Zr_x)O_3$ and $(Pb_{1-x}, La_x)TiO_3$. Solid State Commun. *16*, 931-935 (1975)
290 R. Merlin, J.A. Sanjurjo, A. Pinczuk: "Raman Scattering from Soft Phonons of $Pb(Ti_{1-x},Zr_x)O_3$ and $(Pb_{1-x},La_x)TiO_3$", in Ref.2, pp.895-899
291 M.D. Mermelstein, H.Z. Cummins: Dynamic central peak in ferroelectric KH_2PO_4. Phys. Rev. B*16*, 2177-2183 (1977)
292 R. Migoni, H. Bilz, D. Bauerle: Origin of Raman scattering and ferroelectricity in oxidic perovskites. Phys. Rev. Lett. *37*, 1155-1158 (1976)
293 W. Muller-Lierheim, H.H. Otto: Temperature dependence of LO splitting of A-modes in $Pb_5Ge_3O_{11}$. Solid State Commun. *24*, 349-352 (1977)
294 W. Muller-Lierheim, T. Suski, H.H. Otto: Factor group analysis of the Raman spectrum of $Pb_5Ge_3O_{11}$. Phys. Status Solidi B *80*, 31-41 (1977)
295 R.E. Nettleton: Temperature dependence of tunnel and mixed modes in KH_2PO_4. Ferroelectrics *9*, 59-61 (1975)
296 I.A. Oxton, O. Knop: Infrared evidence of a low-temperature transition in $(NH_4)_2CuCl_4 \cdot 2H_2O$. Chem. Phys. Lett. *49*, 560-562 (1977);
– K.F. Pai, T.J. Parker, R.P. Lowndes: "The Temperature Dependence of the Dielectric Response of Pseudo Displacive Ferroelectrics by Dispersive Fourier Transform Spectroscopy", in *2nd Int. Conf. and Winter School on Submillimeter Waves and their Applications*, San Juan, Puerto Rico, 6-11 Dec 1976, (IEEE, New York 1976) pp. 176 et seq.;
– T.J. Parker, D.A. Ledsham, W.G. Chambers: Study of soft mode behaviour in KDP by dispersive Fourier transform spectroscopy. Ferroelectrics *14*, 625 (1976)
297 P.S. Peercy: Raman scattering near the tricritical point in SbSI. Phys. Rev. Lett. *35*, 1581-1584 (1975)
298 P.S. Peercy: Measurement of the 'Soft' mode and coupled modes in the paraelectric and ferroelectric phases of KH_2PO_4 at high pressure. Phys. Rev. B *12*, 2725-2240 (1975)
299 P.S. Peercy: 'Soft' mode and coupled modes in the ferroelectric phase of KDP. Solid State Commun. *16*, 439-442 (1975)
300 P.S. Peercy: Pressure studies of Raman scattering from soft modes: KH_2PO_4. Comments Solid State Phys. *7*, 37-43 (1975)
301 A.F. Penna, A. Chaves, S.P.S. Porto: Debye-like diffusive central mode near the phase transition in ferroelectric lithium tantalate. Solid State Commun. *19*, 491-494 (1976)
302 A.F. Penna, S.P.S. Porto, A.S. Chaves: "High Temperature Light Scattering in Lithium Tantalate", in Ref. 2, pp. 890-894
303 J. Petzelt, J. Grigas: Far infrared dielectric dispersion in Sb_2S_3, Bi_2S_3 and Sb_2Se_3 single crystals. Ferroelectrics *5*, 59-68 (1973)
304 J. Petzelt, J. Grigas, I. Mayerova: Far infrared properties of the pseudoproper ferroelectric ammonium sulfate. Ferroelectrics *6*, 225-234 (1974)
305 M. Peyrard, J. Garandet, M. Remoissenet: Temperature dependence of far infrared reflectivity of lithium iodate. Solid State Commun. *16*, 227-231 (1975)

306 A. Pinczuk, G. Burns, F.H. Dacol: Soft optic phonon in ferroelastic $BiVO_4$. Solid State Commun. *24*, 163 (1977)
307 L.M. Plyasova, V.I. Zharkov, G.N. Kustova, L.G. Karakchiev, M.M. Andrushkevich: Polymorphism of cobalt molybdate. Inorg. Mater. *9*, 466-468 (1973)
308 H. Poulet, J.P. Mathieu: Raman spectra and phase transitions of ammonium paraperiodate $(NH_4)_2H_3IO_6$. J. Raman Spectrosc. *4*, 231-234 (1976)
309 W. Prettl, K.H. Rieder, R. Nitsche: Raman scattering investigation of ferroelastic Sb_5O_7I crystals. Z. Phys. B *22*, 49-58 (1975)
310 W. Prettl: Raman scattering study and symmetry aspects of the phase transition in ferroelastic Sb_5O_7I. Phys. Rev. B *14*, 2171-2179 (1976)
311 W. Prettl, K.H. Rieder: "Raman Scattering Study of the Ferroelastic Phase Transition in Sb_5O_7I Crystals", in Ref. 2, pp. 939-943
312 W. Prettl, K.H. Rieder, R. Nitsche: Study of the purely ferroelastic phase transition in α-Sb_5O_7I by Raman spectroscopy. Ferroelectrics *13*, 337-339 (1976)
313 M. Quilichini, J.F. Ryan, J.F. Scott, H.J. Guggenheim: Light scattering from soft modes in ferroelectric $BaZnF_4$ and $BaMgF_4$. Solid State Commun. *16*, 471-475 (1975)
314 A.M. Quittet, M.I. Bell, M. Krauzman, P.M. Raccah: Anomalous scattering and asymmetrical line shapes in Raman spectra of orthorhombic $KNbO_3$. Phys. Rev. B *14*, 5068-5072 (1976)
315 L.M. Rabkin, V.P. Dmitriev, V.I. Torgashev, L.A. Shuvalov, N.M. Shagina: Vibrational spectra and particularities of the phase transitions in alkaline trihydrogen selenites. Ferroelectrics *14*, 627-629 (1976)
316 C. Rocchiccioli-Deltcheff: Comparison of infrared absorption spectra of niobates and tantalates of monovalent metals. Spectrochim. Acta A *29a*, 93-106 (1973) (in French)
317 T. Romanovskis, J. Zvirgzde: Light scattering near the critical electric point in barium titanate. Phys. Status Solidi A *32*, K41-K45 (1975)
318 M. Rousseau, J.Y. Gesland, J. Julliard, J. Nouet, J. Zaremswitch, A. Zaremswitch: Crystallographic, elastic and Raman scattering investigations of structural phase transitions in $RbCdF_3$ and $TlCdF_3$. Phys. Rev. B*12*, 1579 (1975)
319 R. Ruppin: Infrared absorption and ferroelectricity of small $BaTiO_3$ crystals. Phys. Status Solidi B *64*, 701-707 (1974)
320 Kh.Sh. Rustamov, V.S. Gorelik, Yu.S. Kuz'minov, G.V. Peregudov, M.M. Suchchinskii: Influence of temperature on the Raman scattering spectra of a $Ba_{0.25}Sr_{0.75}Nb_2O_6$. Sov. Phys.-Solid State *18*, 1988-1990 (1976)
321 J.F. Ryan, J.F. Muratore, J.F. Scott: Light scattering from soft modes in $BaM^{2+}F_4$: a new class of antiferroelectric. Ferroelectrics *7*, 279-281 (1974)
322 E. Sailer, H.G. Unruh: On order parameter fluctuations in Rochelle salt detected by Brillouin scattering. Solid State Commun. *16*, 615-619 (1975)
323 T. Sakudo, H. Uwe: Raman scattering study of the soft phonon modes in the ferroelectric $SrTiO_3$. Ferroelectrics *8*, 587-588 (1974)
324 E. Salje, K. Ishi: Ferroelastic phase transitions in lead phosphate-vanadate $Pb_3(P_xV_{1-x}O_4)_2$. Acta Crystallogr. Sect A A*33*, 3399-3408 (1977)
325 I. Savatinova, E. Anachkova: Raman study of dynamic disorder and the ferroelectric transition in $K_4Fe(CN)_6 \cdot 3H_2O$. Phys. Status Solidi B *82*, 677-683 (1977)
326 A. Sawada, M. Udagawa, T. Nakamure: Proper ferroelastic transition in piezoelectric lithium ammonium tartrate. Phys. Rev. Lett. *39*, 829-832,902 (1977)
327 A. Scalabrin: "Temperature Dependence of the Phonons in $BaTiO_3$ and Some Effects of the Exciton on the Raman Scattering of Cu_2O and CdS", Univ. Southern California, Los Angeles, USA, Ph.D. Thesis, Univ. Microfilms, Ann Arbor, Mich., USA. Order No. 74-26049 (1974) 115p.
328 A. Scalabrin, S.P.S. Porto, H. Vargas: Temperature dependence of the broad A_1(TO) phonon Raman linewidth in $BaTiO_3$. Solid State Commun. *24*, 291-294 (1977)
329 M.A.F. Scarparo, R. Srivastava, R.S. Katiyar: "A Comparative Study of the Raman Spectra of Crystals with the KDP Structure", in Ref. 2, pp. 839-844
330 T. Sekine, K. Uchinokura, E. Matsuura: Raman scattering from two-phonon resonance states in $SrTiO_3$. Solid State Commun. *18*, 569-572 (1976)
331 V.F. Shabanov, A.V. Sorokin, A.P. Fedotov: Low frequency Raman spectra of TGSe. Zh. Prikl. Spektrosk. *21*, 750-752 (1974) (in Russian)

332 V.F. Shabanov, A.V. Sorokin: Investigation into the Raman scattering spectra of lattice vibrations of crystals of the triglycine sulphate family. Bull. Acad. Sci. USSR: Phys. Ser. *39*, (4): 81 (1975)
333 S.M. Shapiro, H.Z. Cummins: Critical opalescence in quartz. Phys. Rev. Lett. *21*, 1578 (1968)
334 C.Y. She, C.L. Pan: Raman scattering of high-temperature phase transition in KH_2PO_4. Solid State Commun. *17*, 529-531 (1975)
335 T. Shigenari, Y. Takagi: Ellects of deuteration on proton tunneling modes and coupled optic modes in the Raman spectrum of KDP crystals. J. Phys. Soc. Jpn. *42*, 1650-1655 (1977)
336 T. Shimada, K.L.I. Kobayashi, Y. Katayama, K.F. Komatsubara: Soft-phonon-induced Raman scattering in IV-VI compounds. Phys. Rev. Lett. *39*, 143-146 (1977)
337 T.Y. Shon, S.S. Mitra: Order-disorder phenomenon in sodium nitrate studied by low-frequency Raman scattering. Phys. Rev. B *12*, 4530-4533 (1975)
338 D.F. Shriver, G. Joy: Raman line broadening in the superionic conductors, $Ag_2(HgI_4)$, and $Cu_2(HgI_4)$. J. Electrochem. Soc. *123*, 588-590 (1976)
339 D. Siapkas: Far-infrared reflectivity spectra of BiSi. Ferroelectrics *7*, 295-296 (1974)
340 G.J. Simonis, C.E. Hathaway: Raman spectrum and phase transition in sodium azide. Phys. Rev. B *10*, 4419-4433 (1974)
341 G.A. Sindenskii, N.N. Krainik, V.A. Trepakov, N.B. Nazarenko, M.S. Sactykov, V.A. Pir'mennyi, K.P. Skornyakova: Investigation into light scattering in a $PbMg_{0.5}Nb_{0.67}O_3$ crystal at diffuse ferroelectric phase transition temperatures. Bull. Acad. Sci. USSR: Ser. Phys. *39*, (4): 129 (1975)
342 G.A. Smolenskii, V.A. Trepakov, N.N. Krainik: Light scattering in lead magnonio-bate in the ferroelectric phase transition temperature range. JETP Lett. *20*, 143 (1974)
343 E.F. Steigmeier, H. Auderset, G. Harbeke: The central peak in SbSI. Phys. Status Solidi B *70*, 705-716 (1975)
344 B.A. Strukov, K.A. Minaeva, V.I. Teleshevskii, N.G. Shirina, S.K. Kkhanna: Critical anomalies in the velocity of ultrasound in single crystal of triglycine sulphate. Bull. Acad. Sci. USSR: Phys. Ser. *39*, (4): 101 (1975)
345 S. Sugai, K. Murase, H. Kawamura: Observation of soft TO-phonon in SnTe by Raman scattering. Solid State Commun. *23*, 127-129 (1977)
346 H.C. Tang, B.H. Torrie: Raman study of NH_4NO_3 and ND_4NO_3-250-420K. J. Phys. Chem. Solids *38*, 125-138 (1977)
347 W. Taylor, D.J. Lockwood, J.W. Arthur, T.J. Hosea: The Raman spectrum of lead germanate. Ferroelectrics *12*, 113-115 (1976)
348 M.K. Teng, M.N. Spilbauer, M. Balkanski: Resonance and exchange effects in Raman intensity in SbSI. J. Phys. 36, 153-158 (1975) (in French)
349 M.K. Teng, M. Massot, M. Balkanski: "Influence of the Substitution of Sb by Bi on the Soft Mode Behavior in $Bi_xSb_{1-x}SI$", in Ref. 2, pp. 806-811
350 J.C. Toledano, G. Errandonea, J.P. Jaguin: Soft acoustic mode in ferroelastic lanthanum pentaphosphate. Solid State Commun. *20*, 905-907 (1976)
351 J.C. Toledano, M. Busch, J. Schneck. Mechanism of the ferroelastic transition in barium-sodium niobate. Ferroelectrics *13*, 327-328 (1976)
352 H.-G. Unruh, O. Ayere: Raman and dielectric measurements on ferrielectric ammonium sulphate. Ferroelectrics *12*, 181-183 (1976)
353 S. Ushioda, D. Heiman, J.P. Remeika: "Frequency Dependent Damping Function of the Soft Mode in $PbTiO_3$", in Ref. 2, pp. 457-461
354 H. Uwe, T. Sakudo: Stress-induced ferroelectricitiy and soft phonon modes in $SrTiO_3$. Phys. Rev. B *13*, 271-286 (1976)
355 H. Uwe, T. Sakudo: Raman-scattering study of stress-induced ferroelectricity in $KTaO_3$. Phys. Rev. B *15*, 337-345 (1977)
356 A.S. Vasilevskaya, I.M. Grodnenskii, A.S. Sonin: Phase transitions in a (Pb, La)(Zr, Ti)O_3 ferroelectric solid solution. Sov. Phys.-Solid State *18*, 2165-2166 (1976)
357 J.L. Verble: "Raman Scattering in the Ferroelectric Phase of Strontium Barium Niobate: Comparisons between SBN and Nd:SBN", in Ref. 2, pp. 900-904
358 J.E. Vesel, B.H. Torrie: A Raman scattering study of the solid phases of HCl, DCl, HBr, and DBr. Can. J. Phys. *55*, 592-603 (1977) (in English)

359 K. Viswanathan, E. Salje: Thermal contraction in NH_4IO_3 and its significance with respect to the phase transition. Acta Crystallogr. A *A31*, 810-813 (1975)
360 C.H. Wang, M. Tatsumi, T. Matsuo: Raman scattering study of the protonic order-disorder transition in $SnCl_2 \cdot 2H_2O$. J. Chem. Phys. *67*, 3097-3105 (1977)
361 V. Winterfeldt, G. Schaack, A. Klopperpieper: Temperature behaviour of optical phonons near T_c in triglycine sulphate and triglycine selenate. I. Infrared reflection and Raman spectra. Ferroelectrics *15*, 21-34 (1977)
362 Y. Yacoby, S. Yust: Differential Raman scattering from soft modes in mixed crystals of $K_{1-x}Na_xTaO_3$ and $K_{1-x}Li_xTaO_3$. Ferroelectrics *7*, 271-273 (1974)
363 T. Yagi, M. Tokunaga, I. Tatsuzaki: Brillouin scattering studies of the uniaxial ferroelectric phase transition in TGS and TGSe. J. Phys. Soc. Jpn. *40*, 1659-1667 (1976)
364 G.N. Zhizhin, T.P. Myasnikova, V.N. Rogovoi: Infrared reflection spectra of ammonium sulphate. Sov. Phys.-Solid State *17*, 821-822 (1975)
365 A.I. Zvyagin, S.D. El'chaninova, T.S. Stetsenko, L.N. Pelikh, E.N. Khatsko: A low-temperature structural phase transition in cesium-dysprosium molybdate (from infrared spectra and magnetic susceptibility observations). Fiz. Nizk. Temp. *1*, 79-82 (1975) (in Russian)

7 *Zone Center Transitions: Other Optical Techniques*

366 M. Adam, G.M. Searby: Optical studies of NH_4Cl near -30 C. Phys. Status Solidi A *19*, 185-192 (1973)
367 A.T. Anistratov, A.V. Mel'nikova: Optical and electrooptical properties of crystals from the $Me^{+}NH_4RO_4.NH_2O$ group. Bull. Acad. Sci. USSR: Phys. Ser. *39*, (4): 144 (1975)
368 J.P. Bachheimer, G. Dolino: Measurement of the order parameter of α-quartz by second harmonic generation of light. Phys. Rev. B *11*, 3195-3205 (1975)
369 J.G. Bergman: Molecular mechanics of the ferroelectric to paraelectric phase transition in $LiNbO_3$ via optical second harmonic generation. Chem. Phys. Lett. *38*, 230-233 (1976)
370 J.G. Bergman: Molecular mechanics of the ferroelectric to paraelectric phase transition in $LiTaO_3$ via optical second harmonic generation. J. Chem. Soc. *98*, 1054-1055 (1976)
371 J.G. Bergman, G.R. Crane: Structural aspects of nonlinear optics-tetrahedral rotations. Chem. Phys. Lett. *41*, 133-136 (1976)
372 J.G. Bergman: Molecular mechanics of the ferroelectric to paraelectric phase transition in $Ba_6Ti_2Nb_8O_{30}$ via second harmonic generation. Opt. Comm. *18*, 51-52 (1976)
373 J.G. Bergman, G.R. Crane, E.H. Turner: The tetragonal deformation of the TiO_6 octahedron in ferroelectric $PbTiO_3$. J. Solid State Chem. *21*, 127-133 (1977)
374 D. Bernard, J. Lucas, L. Rivoallan: The ferroelectric pyrochlore derivatives of $Cd_2Nb_2O_6S$. An investigation of phase transitions by the techniques of nonlinear optics. Solid State Commun. *18*, 927-930 (1976) (in French)
375 H.J. Bernhardt: Optical measurements of the phase transition point in KDP-type crystals by domain observation. Exp. Tech. Phys. *25*, 19-25, (1977)
376 S.T. Buljan, H.A. Mckinstry, V.S. Stubican: Optical and X-ray single crystal studies of the monoclinic-tetragonal transition in ZrO_2. J. Am. Ceram. Soc. *59*, 351-354 (1976);
– T.S. Chang: "Nonlinear Optical and Lattice Dynamical Studies of Some Ferroelectric Crystals", Stanford Univ., Calif, USA, Ph.D. Thesis, Univ. Microfilms, Ann Arbor, Mich., USA Order No. 72-16704 (1972) 168p
377 G. Dolino: Effects of domain shapes on second harmonic scattering in TGS. Phys. Rev. B *6*, 4025 (1972)
378 I. Freund: Long range order in NH_4Cl. Phys. Rev. Lett. *24*, 1017 (1970)
379 H. Iwasaki, S. Miyazawa, H. Koizumi, S. Sugii, N. Niizeki: Ferroelectric and optical properties of $Pb_5Ge_3O_{11}$ and its isomorphous compound $Pb_5Ge_2SiO_{11}$ (and thermal expansion). J. Appl. Phys. *43*, 4907-4915 (1972);
– J. Kobayashi: Peculiar physical properties of improper ferroelectrics. Ferroelectrics *10*, 277-282 (1976)

380 M.P. Lisitsa, U. Nasirov, I.D. Turjanitsa, I.V. Fekeshgazi: "Nonlinear Light Absorption in Crystalline and Vitreous SbSI", in *Proc. of the Int. Conf. on Amorphous Semiconductors*, Balatonfured, Hungary, 20-25 Sept. 1976, ed. by I.K. Somogyi (Akad. Kiado, 1976) pp. 333-336
381 G.G. Mitin, V.S. Gorelik, B.N. Matsonashvili, M.M. Sushchinskii: Study of the phase transitions in barium titanate by the optical second harmonic method. Sov. Phys.-Solid State *16*, 817-818 (1974)
382 T.V. Panchenko, Yu. I. Samchenko, V.G. Taran, S.A. Flerova: Emission of light during the repolarization of $BaTiO_3$ single crystals. Sov. Phys.-Crystallogr. *21*, 481-482 (1976)
383 J. Ravez, A. Perron, P. Chaminade, P. Hagenmuller, L. Rivoallan: Studies of the crystallographic, dielectric and linear optical properties of some new phases of composition $Ba_xLi_5Nb_{5(1-y)}Ta_{5y}O_{15}$ and the oxidized tetragonal tungsten bronze structure. J. Solid State Chem. *10*, 274-281 (1974) (in French)
384 M.J. Ravez, A. Perron-Simon, B. Elouadi, L. Rivoallan, P. Hagenmuller: Evolution of some physical properties of phases of 'quadratic tungsten bronze' structure by substitution of the alkaline-earth element by lead. J. Phys. Chem. Solids *37*, 949-952 (1976) (in French)
385 N.A. Romanyuk, A.M. Kostetskii: Dispersion and temperature dependences of refractive indices of Rochelle salt crystals. Sov. Phys.-Solid State *18*, 867-869 (1976)
386 V.K. Semenchenko, I.T. Bodnar: Investigation of the refractive indices of quartz in the α-β transition region. Opt. Spectrosc. *35*, 56-58 (1973)
387 A.W. Sleight, J.D. Bierlein: Phase transition in $Cd_2Ta_2O_7$ and the $Cd_2Nb_{2-x}Ta_xO_7$ series. Solid State Commun. *18*, 163-166 (1976)
388 G. Stuhmer, E. Rieflin: Ultraviolet reflectivity of NH_4Cl and NH_4I in ordered and disordered phases. Opt. Commun. *9*, 103-105 (1973)
389 M. Takashige, S. Hirotsu, S. Sawada: Optical rotatory power and 60 C phase transition of $Ca_2Pb(C_2H_5CO_2)_6$. J. Phys. Soc. Jpn. *38*, 904 (1975)
390 M. Vallade: Simultaneous measurements of the second harmonic generation and the birefringence of KH_2PO_4 near its ferroelectric point. Phys. Rev. B *12*, 3755-3765 (1975)
391 Z.S. Vasilina, N.A. Romanyuk: Fundamental reflectance spectra of triglycine sulfate. Opt. Spectrosc. *35*, 177-178 (1973)
392 K.A. Verkhovskaya, V.N. Nosov, V.M. Fridkin: Spectral sensitization during phase transition. Sov. Phys.-Solid State *17*, 2025-2026 (1975);
– R. Verreault: Crystallographic, optical and magnetic properties of Eu_2SiO_4. Phys. Kondens. Mater. *14*, 37-54 (1971);
– H. Vogt: Study of structural phase transitions by techniques of nonlinear optics. Appl. Phys. *5*, 85-96 (1974)
393 T. Yamada, H. Iwasaki: Ferroelectric, piezoelectric, and optical properties of $SrTiO_3$ single crystals and phase-transition points in the solid-solution systems. J. Appl. Phys. *44*, 3934-3939 (1973)
394 V.K. Yanovskii, V.I. Voronkova, A.L. Aleksandrovskii, V.A. D'yakov: Structure and properties of Bi_2WO_6 ferroelectric crystals. Sov. Phys.-Dokl. *20*, 306-307 (1975)

8 *Zone Boundary Transitions: Birefringence*

395 K.S. Aleksandrov, A.T. Anistratov, A.I. Krupnyi, L.A. Pozdnyakova, S.V. Mel'nikova, B.V. Beznosikov: X-ray, optical and ultrasonic investigations of phase transitions in $TlMnCl_3$. Sov. Phys.-Solid State *17*, 471-473 (1975)
396 A.T. Anistratov, V.G. Martynov, S.V. Mel'nikova: Optical properties of gadolinium molybdate near an improper ferroelectric transition. Sov. Phys.-Solid State *17*, 1964-1966 (1975);
– P. Bastie, J. Bornarel, J.F. Legrand: Perturbated regions in the vicinity of domain wall in ferroelastic-ferroelectric crystals. Ferroelectrics *13*, 455-458 (1976)
397 J.B. Bates, R.W. Major, F.A. Modine: Phase transitions in $RbCaF_3$. I. Optical studies. Solid State Commun. *17*, 1347-1350 (1975)

398 A.A. Berezhnoi, I.A. Velichko, Yu. V. Popov: Switching dynamics of spontaneous birefringence in GMO. Opt. Spectrosc. *42*, 655 (1977);
– E. Courtens: "The Observation of Birefringence in Fluctuating Anisotropic Media. Application to the 105K Phase Transition in $SrTiO_3$", in Ref. 25, pp. 293-311
399 S.N. Drozhdin, B.G. Bochkov, N.D. Gavrilova, T.V. Popova, V.A. Koptsik, V.K. Novik: Ferroelectric properties of Cu-Br- and Co-I-boracites. Sov. Phys. -Crystallogr. *20*, 526-527 (1975)
400 J. Fousek, M. Glogarova: Elastooptic effect and sources of the spontaneous birefringence in $Gd_2(MoO_4)_3$. Solid State Commun. *17*, 97-100 (1975)
401 G. Heygster, W. Kleemann: Optical investigations on magnetic and structural phase transitions of $(CH_3NH_3)_2CuCl_4$ and $(C_2H_5NH_3)_2CuCl_4$. Physica B C *89c*, 165-176 (1977)
402 S. Hirotsu, S. Sawada: Spontaneous Birefringence and order parameter of $KMnF_3$ below the 186K transition point. Solid State Commun. *12*, 1003-1005 (1973)
403 S. Hirotsu: Some optical and thermal properties of $CsCuCl_3$ and its phase transition near 423K. J. Phys. C *8*, L12-L16 (1975)
404 N.A. Irisova, G.V. Kozlov, T.N. Narytnik, I.M. Chernyshev: Temperature dependence of the electrooptical effect of single $SrTiO_3$ crystals in the submillimetre range. Izv. Akad. Nauk SSSR Ser. Fiz. 39, 795-797 (1975);
– I.R. Jahn, I.H. Brunskill, H. Dacts, R. Bausch: The β,γ phase transition in the system $NH_4Cl_{1-x}Br_x$. J. Phys. C *8*, 3280-3286 (1975);
– I.R. Jahn, I.H. Brunskill: Nature of the β,γ phase transition in $NH_4Cl_{1-x}Br_x$ mixed crystals. Acta Crystallogr. A A*31*, S189 (1975)
405 K. Knorr, I.R. Jahn, G. Heger: Birefringence, X-ray and neutron diffraction measurements on the structural phase transitions of $(CH_3NH_3)_2MnCl_4$ and $(CH_3NH_3)_2FeCl_4$. Solid State Commun. *15*, 231-238 (1974)
406 S. Kojima, K. Ohi, T. Nakamura: Temperature dependence of electrooptic coefficient and spontaneous birefringence of ferroelectric gadolinium molybdate. J. Phys. Soc. Jpn. *41*, 162-166 (1976)
407 F.A. Modine, E. Sonder, W.P. Unruh, C.B. Finch, R.D. Westbrook: Phase transitions in $RbCaF_3$. Phys. Rev. B *10*, 1623-1634 (1974)
408 D. Semmingsen, J. Feder: A structural phase transition in squaric acid. Solid State Commun. *15*, 1369-1372 (1974)
409 F. Smutny, C. Konak: Dielectric, optical, and electrooptical properties of cobalt-iodine boracite ($Co_3B_2O_{13}I$). Phys. Status Solidi A *31*, 151-158 (1975)

9 *Zone Boundary Transitions: Optical Absorption and Fluorescence*

410 V.G. Bochkov, V.I. Bugakov, K.A. Verkhovskaya, T.M. Polkhovskaya, V.M. Fridkin: Fundamental absorption in ferroelectric boracites and gadolinium molybdate. Sov. Phys.-Solid State *16*, 1217-1219 (1975)
411 W. Bohm, R. Herb, H.G. Kahle, A. Kasten, J. Laugsch, W. Wuchner: Spectroscopic confirmation of a crystallographic phase transition and of antiferromagnetic ordering in $TbPO_4$. Phys. Status Solidi B *54*, 527-536 (1972);
– E.V. Bursian, Ya. G. Girshberg, A.V. Ruzhnikov: The correlation between optical absorption spectra, carrier mobility, and phase transition temperature in some ferroelectrics. Phys. Status Solidi B *74*, 689-693 (1976)
412 M. Faucher, P. Caro: Optical study of $LaAlO_3$: Eu at temperatures approaching the rhombohedric-cubic transition. J. Chem. Phys. *63*, 446-454 (1975)
413 V.M. Fridkin, K.A. Verkovskaya, B.G. Bochkov: Optical absorption edge in improper ferroelectrics in the phase transition region. Phys. Status Solidi A *22*, 759-766 (1974)
414 R. Laiho, M. Natarajan, M. Kaira: Some electrical and optical studies in $CsCuCl_3$ crystals. Phys. Status Solidi A *15*, 311-317 (1973)
415 N.N. Nesterova, R.V. Pisarev, G.T. Andreeva: Optical absorption of ferroelectrical copper and chromium boracites. Phys. Status Solidi B *65*, 103-110 (1974);
– M.D. Sturge, E. Cohen, L.G. van Uitert, R.P. van Stapele: Microscopic order parameters in $PrAlO_3$. Phys. Rev. B *11*, 4768-4779 (1975)

416 P. Alain, B. Piriou: High temperature Raman scattering and phase transition in $EuAlO_3$. Solid State Commun. *17*, 35-39 (1975)
417 C.H. Barta, A.A. Kaplyanskii, V.V. Kulakov, Yu.F. Markov: Soft mode of the Brillouin zone boundary and the character of the phase transition in monovalent mercury halide crystals. JETP Lett. *21*, 54 (1975)
418 C. Barta, A.A. Kaplyanskii, V.V. Kulakov, Yu. F. Markov: Observation of a phase transition in calomel (Hg_2Cl_2) crystals. Sov. Phys.-Solid State *16*, 2022 (1975)
419 C. Barta, J.P. Chapelle, G. Hauret, A.N. Cao-Xuan, A. Fouskova, C. Konak: Brillouin scattering in Hg_2Cl_2. Phys. Status Solidi A *34*, K51-K54 (1976);
– J.B. Bates, R.W. Major, F.A. Modine: Phase transitions in $RbCaF_3$. I. Optical studies. Solid State Commun. *17*, 1347-1350 (1975)
420 D. Bauerle, A. Pinczuk: Low frequency vibrational modes and the phase transitions of rhombohedral $PbTi_{1-x}Zr_xO_3$. Solid State Commun. *19*, 1169-1171 (1976)
421 D. Bauerle, W.B. Holzapfel, A. Pinczuk, Y. Yacoby: Temperature and hydrostatic pressure dependence of vibrational modes in $PbTi_{1-x}Zr_xO_3$. Phys. Status Solidi B*83*, 99-107 (1977);
– W. Bauhofer, L. Genzel, C.H. Perry, I.R. Jahn: "Optical Phonons and Phase Transitions in $(NH_4)Cl_{1-x}Br_x$ Mixed Crystals", in Ref. 2, pp. 918-922
422 J. Berger: The Raman spectra and the antiferroelectric phase transition of copper formate tetrahydrate. J. Phys. C *8*, 2903-2910 (1975)
423 J. Berger, G. Hauret, M. Rousseau: Brillouin scattering investigation of the structural phase transition of $TlCdF_3$. Solid State Commun. *25*, 569-571 (1978)
424 A. Bree, M. Edelson: A study of the second order phase transition in biphenyl at 40K through Raman spectroscopy. Chem. Phys. Lett. *46*, 500-504 (1977)
425 W.J.R. Burke, R.J. Pressley, J.C. Slonczewski: Raman scattering in $SrTiO_3$. Solid State Commun. *9*, 121, 191 (1971)
426 R.P. Canterford, F. Ninio: The Raman spectrum of copper formate tetrahydrate. II. J. Phys. C *8*, 385-388 (1975)
427 A.N. Cao-Xuan, G. Hauret, J.P. Chapelle: Brillouin scattering in Hg_2Cl_2. Solid State Commun. *24*, 443-445 (1977)
428 J.C. Castro, H.C. Basso, M. de Souza: Local mode of H^- substitutional ion in KCN. Phys. Status Solidi B *77*, 685-692 (1976)
429 J.M. Courdille, R. Deroche, J. Dumas: Critical behavior of acoustical waves in ferroelectric-ferroelastic phase of $Tb_2(MoO_4)_3$. J. Phys. *36*, 891-895 (1975)
430 W. Dultz: Critical effects of the light scattering intensity at an order-disorder phase transition: KCN. J. Chem. Phys. *65*, 2812-2816 (1976)
431 L.A. Firstein, G.A. Barbosa, S.P.S. Porto: "Combined Raman-Brillouin Scattering in $SrTiO_3$ Near 100K Phase Transition", in Ref. 2, pp. 866-871
432 P.A. Fleury: Anomalous phonon behavior near the phase transition in ferroelastic-ferroelectric GMO. Solid State Commun. *8*, 601-605 (1970)
433 D. Fontaine, H. Poulet: Influence of phase transitions on scattering and absorption spectra of KCN. Phys. Status Solidi B *58*, K9-K12 (1973) (in French)
434 B.N. Ganguly, F.G. Ullman, R.D. Kirby, J.R. Hardy: Effect of uniaxial stress on the unstable A_1 phonon in ferroelectric GMO. Phys. Rev. B*12*, 3783-3788 (1975)
435 B.N. Ganguly, F.G. Ullman, R.D. Kirby, J.R. Hardy: Optic-mode coupling in ferroelectric gadolinium molybdate. Solid State Commun. *17*, 533-536 (1975)
436 B. Ganguly: "Effect of Temperature and Pressure on Raman Scattering in Gadolinium Molybdate"; Univ. Nebraska, Lincoln, USA, Ph.D. Thesis, Univ. Microfilms, Ann Arbor, Mich., USA. Order No. 75-3422 (1975) 143 p
437 B.N. Ganguly, F.G. Ullman, R.D. Kirby, J.R. Hardy: Raman spectrum of gadolinium molybdate at 80K. Phys. Rev. B *13*, 1344-1352 (1976)
438 B.N. Ganguly, F.G. Ullman, R.D. Kirby: Comment on Raman scattering from soft modes in GMO. J. Phys. Soc. Jpn. *43*, 1085 (1977)
439 G. Gautier, M. Debeau: Laser-Raman spectrum of monoclinic β sulphur. Spectrochim. Acta A *32a*, 1007-1010 (1976)
440 D.M. Hanson: Direct observation of a soft phonon associated with a structural phase transition in a molecular crystal. J. Chem. Phys. *63*, 5046-5047 (1975);
– G. Heygster, W. Kleemann: Optical investigations on magnetic and structural phase transitions of $(CH_3NH_3)_2CuCl_4$ and $(C_2H_5NH_3)_2CuCl_4$. Physica B C *89c*, 165-176 (1977);

– H.D. Hochheimer, E. Spanner, D. Strauch: Phase diagram of ammonium iodide obtained by Raman spectroscopy under hydrostatic pressure. J. Chem. Phys. *64*, 1583-1585 (1976)
441 H.D. Hochheimer, T. Geisel: Study of phase transitions and of the possible existence of a fifth phase in NH_4Br by Raman scattering under high pressure. J. Chem. Phys. *64*, 1586-1592 (1976)
442 H.D. Hochheimer, W.F. Love, C.T. Walker: High-pressure Brillouin scattering study of the order-disorder transition in KCN. Phys. Rev. Lett. *38*, 832-835 (1977)
443 V.T. Hochli: Elastic constants and soft optical modes in GMO. Phys. Rev. B *6*, 1814-1823 (1972)
444 Z. Iqbal, L.H. Sarma, K.D. Moller: Infrared and Raman spectrum of KNCS crystal. Mode analysis and the order-disorder phase transition. J. Chem. Phys. *57*, 4728-4737 (1972)
445 Z. Iqbal, C.W. Christoe: Librational 'soft' mode around the 240K phase transition in thallium azide (TlN_3). Chem. Phys. Lett. *29*, 623-626 (1974);
– S. Itoh, T. Nakamura: Brillouin scattering study of $Gd_2(MoO_4)_3$ using a double Fabry-Perot interferometer. Jpn. J. Appl. Phys. *14*, (Suppl) 183-186 (1975)
446 T.E. Jenkins, L.T.H. Ferris, A.R. Bates: Critical behavior of the light scattering intensity at the phase transition in hexamine cadmium (III) halides. J. Phys. C *10*, L521-L526 (1977)
447 A.A. Kaplyanskii, V.V. Kulakov, Yu.F. Markov, C. Barta: The soft mode properties in Raman spectra of improper ferroelastics Hg_2Cl_2 and Hg_2Br_2. Solid State Commun. *21*, 1023-1025 (1977)
448 A.S. Knyazev, Yu.M. Poplavko, V.P. Zakharov, L.G. Kosakovskii: Vibrational spectra of antiferroelectrics. Sov. Phys.-Solid State *16*, 1808-1809 (1974)
449 B.B. Lavrencic, I. Levstek, M. Copic, S. Trost: Dynamics of α-β transition in $NaH_3(SeO_3)_2$: Brillouin scattering study. Ferroelectrics *14*, 637-639 (1976)
450 B.B. Lavrencic, I. Levstek, M. Copic, S. Trost: "Brillouin Scattering in $NaH_3(SeO_3)_2$", in Ref. 2, pp. 928-931
451 D.J. Lockwood: Isolating the totally symmetric Raman spectrum of cubic crystals: the A_1 spectrum of $Cr_3B_7O_{13}Cl$. J. Raman Spectrosc. *2*, 555-562 (1974)
452 D.J. Lockwood: "A Raman Study of the Ferroelectric Phase Transition in Chromium Chlorine Boracite", in Ref. 2, pp. 933-938
453 D.J. Lockwood: Raman spectral study of the ferroelectric phase transition in boracites. Ferroelectrics *13*, 353-354 (1976)
454 Y. Luspin, G. Hauret: Brillouin scattering in gadolinium molybdate near the transition point. J. Phys. Lett. *35*, L193-L195 (1974) (in French)
455 Y. Luspin, G. Hauret: Study of the velocity and damping of acoustic waves obtained by Brillouin scattering in the paraelastic phase in GMO. Phys. Status Solidi B *76*, 551-558 (1976)
456 K.B. Lyons, P.A. Fleury: Phonon interactions and the dynamic central peak in $SrTiO_3$ near the structural phase transition. Solid State Commun. *23*, 477-480 (1977)
457 G.A. Mackenzie, J.W. Arthur, G.S. Pawley: The structural phase transition in octafluoronaphthalene. J. Phys. C *10*, 1133-1149 (1977)
458 J.J. Martin, G.S. Dixon, P.P. Velasco: Phonon scattering in $RbCaF_3$ and $KMnF_3$. Phys. Rev. B *14*, 2609-2612 (1976)
459 J. Petzelt: New type of far IR soft mode in ferroelectric GMO. Solid State Commun. *9*, 1485-1488 (1971)
460 J. Petzelt, I. Mayerova: Far infrared reflectivity of Ni-I and Co-I boracites. Czech. J. Phys. B *B23*, 1277-1280 (1973)
461 J. Petzelt, J. Roos, H. Granicher: The antiferroelectric phase transition in $Ag_2H_3IO_6$. Ferroelectrics *13*, 437-438 (1976)
462 Yu. A. Popkov, S.V. Petrov, A.P. Mokhir: Phase-transition anomalies in the Raman-Mandelstam scattering spectra in $BaMnF_4$ and $BaCoF_4$. Fiz. Nizk. Temp. *1*, 189-192 (1975) (in Russian)
463 H. Poulet, J.P. Mathieu: Polarized Raman study of lattice modes and the 10 C phase transition of β-lithium ammonium sulphate. Solid State Commun. *21*, 421-424 (1977);

– C. Rocchiccioli-Deltcheff: Comparison of infrared absorption spectra of niobates and tantalates of monovalent metals. Spectrochim. Acta A *29a*, 93-106 (1973) (in French)
464 A.J. Rushworth, J.F. Ryan: Raman scattering study of phase transitions in $RbCaF_3$. Solid State Commun. *18*. 1239-1241 (1976);
– J.F. Ryan, J.F. Muratore, J.F. Scott: Light scattering from soft modes in $BaM^{2+}F_4$: a new class of antiferroelectric. Ferroelectrics *7*, 279-281 (1974)
465 E. Salje: Structural phase transitions in the system WO_3-$NaWO_3$. Ferroelectrics *12*, 215-217 (1976)
466 J. Sapriel, R. Vacher: Photoelastic tensor components of $Gd_2(MoO_4)_3$. J. Appl. Phys. *48*, 1191-1194 (1977)
467 V.F. Shabanov, A.P. Fedotov: Low-frequency spectrum and first phase transition in $NaH_3(SeO_3)_2$. Sov. Phys.-Solid State *17*, 347-348 (1975)
468 V.F. Shabanov, A.P. Fedotov: Investigation of a phase transition in $NaD_3(SeO_3)_2$ by Raman scattering of light. Sov. Phys.-Solid State *18*, 877-879 (1976)
469 I.W. Shepherd: Wavevector dependent relaxation of an optical phonon in GMO. Solid State Commun. *9*, 1857-1860 (1971)
470 T. Shigenari, Y. Takagi, Y. Wakabayashi: Raman scattering from soft modes in the ferroelectric phase of $Gd_2(MoO_4)_3$. Solid State Commun. *18*,1271-1273 (1976)
471 E.F. Steigmeier, H. Auderset, G. Harbeke: "Soft Mode and Critical Opalescence in $SrTiO_3$", in Ref. 32, pp. 153-158
472 J.H.M. Stoelinga, P. Wyder: Soft mode behaviour and anharmonicity of some two-dimensional layer compounds. J. Chem. Phys. *64*, 4612-4615 (1976)
473 B.I. Swanson, R.R. Ryan: "The Phonon Driven Phase Change in $Cs_2LiCr(CN)_6$", in Ref. 2, pp. 944-947
474 B.H. Torrie, D.J. Lockwood: Raman spectral study of $KMnF_3$. Ferroelectrics *8*, 583-584 (1974)
475 F.G. Ullman, B.N. Ganguly, J.R. Hardy, R.D. Kirby: "Raman Scattering Studies of the Ferroelectric Transition in Gadolinium Molybdate", in Ref. 2, pp. 948-952;
– H. Uwe, T. Sakudo: Stress-induced ferroelectricity and soft phonon modes in $SrTiO_3$. Phys. Rev. B *13*, 271-286 (1976)
476 R.E. Vandenberghe, G.G. Robbrecht, D. Scheerlinck, E. Legrand, V.A.M. Brabers: On the crystallographic properties of some Cu-Mn spinels with 1:3 octahedral superstructure. Acta Crystallogr. A A *31*, S86 (1975)
477 J. Winter, K. Rossler, J. Bolz, J. Pelzl: Mossbauer effect and Raman studies of the structural phase transitions in $K_2(SnCl_6)$. Phys. Status Solidi B *74*, 193-198 (1976)
478 J.M. Worlock, D.H. Olson: "Light Scattering Studies of the Soft Modes Near the Cubic to Tetragonal Phase Transition in $SrTiO_3$", in Ref. 1, pp. 410-414
479 Y. Yacoby, W.W. Kruhler, S. Just: Study of soft modes by temperature-derivative first-and second-order Raman spectroscopy. Phys. Rev. B *13*, 4132-4140 (1976)
480 Y. Yacoby, S. Yust, W.W. Kruhler: Study of soft modes by temperature derivative Raman spectroscopy ($SrTiO_3$). Ferroelectrics *12*, 117-119 (1976)
481 T. Yagi, H. Tanaka, I. Tatsuzaki: Central peak of $KH_3(SeO_3)_2$ studied by light scattering. Phys. Rev. Lett. *38*, 609-612 (1972)
482 T. Yagi, H. Tanaka, I. Tatsuzaki: A soft acoustic mode and a central peak in $KH_3(SeO_3)_2$. J. Phys. Soc. Jpn. *41*, 717-718 (1976)

11 *Zone Boundary Transitions: Other Optical Techniques*

483 F. Borsa, D.J. Benard, W.C. Walker, A. Baviera: NMR and birefringence study of structural transitions in disordered drystals' $Rb_xK_{1-x}MnF_3$. Phys. Rev. B *15*, 84-94 (1977);
– G. Heygster, W. Kleemann: Optical investigations on magnetic and structural phase transitions of $(CH_3NH_3)_2CuCl_4$ and $(C_2H_5NH_3)_2CuCl_4$. Physica B C *89c*, 165-176 (1977);
– S. Hirotsu: Some optical and thermal properties of $CsCuCl_3$ and its phase transition near 423K. J. Phys. C *8*, L12-L16 (1975)
484 A. Sawada, Y. Ishibashi, Y. Takagi: Ferroelasticity and the origin of optical activity of $Ca_2Sr(C_2H_5CO_2)_6$ (DSP). J. Phys. Soc. Jpn. *43*, 195-203 (1977)

485 Yu.V. Shaldin, D.A. Belogurov, T.M. Prokhortseva: Anisotropy of the nonlinear refractive index of single crystals of gadolinium and terbium molybdates. Sov. Phys.-Solid State *15*, 936-938 (1973);
— H. Vogt: Study of structural phase transitions by techniques of nonlinear optics. Appl. Phys. *5*, 85-96 (1974)

12 *Incommensurate Transitions: Birefringence*

486 A.T. Anistratov, S.V. Mel'nikova: Termooptical and dielectric properties of $(NH_4)_2BeF_4$ in the neighborhood of the ferroelectric transition. Sov. Phys.-Crystallogr. *18*, 811-812 (1974)

13 *Incommensurate Transitions: IR Absorption and Light Scattering*

487 K.S. Aleksandrov, A.T. Anistratov, A.I. Krupnyi, V.G. Martynov, Yu.A. Popkov, V.I. Fomin: Mandelstam-Brillouin and ultrasonic studies of phase transitions in the $(NH_4)_2BeF_4$ crystal. Sov. Phys.-Crystallogr. *21*, 296-299 (1976)
488 D.W. Bechtle, J.F. Scott: Anomalous acoustic phonon dispersion in noncommensurate $BaMnF_4$. J. Phys. C *10*, L209-L211 (1977)
489 F. Brehat, J. Claudel, P. Strimer, A. Hadni: Far infrared reflection spectra of the thiourea single crystal. Soft mode study. J. Phys. Lett. *37*, L229-L231 (1976) (in French)
490 E. Castellucci, L.A. Firstein, S.P.S. Porto: "Low Frequency Raman Spectra in the Ferroelectric Antiferroelectric and Paraelectric Phases of $NaNO_2$", in Ref. 2, pp. 872-876
491 J.P. Chapelle, J.P. Benoit: Raman study of external frequencies of thiourea with temperature. J. Phys. C *10*, 145-151 (1977)
492 A. Delahaigue, B. Khelifa, P. Jouve: Infrared Absorption spectra of a single crystal of thiourea in the paraelectric and ferroelectric phases. J. Phys. 33, 507-512 (1972) (in French)
493 A. Delahaigue, B. Khelifa, P. Jouve: Soft mode in thiourea by Raman measurements. Phys. Status Solidi B *72*, 585-589 (1975)
494 V. Fawcett, R.J.B. Hall, D.A. Long, V.N. Sankaranarayanan: A Raman spectroscopy study of the ferroelectric phase transition in K_2SeO_4. J. Raman Spectrosc. *2*, 629-633 (1974)
495 V. Fawcett, R.J.B. Hall, D.A. Long. V.N. Sankaranarayanan: A Raman spectroscopic investigation of a single crystal of potassium selenate, K_2SeO_4 over the temperature range 293-87K. J. Raman Spectrosc. *3*, 229-238 (1975)
496 V. Fawcett, D.A. Long: Raman spectroscopic study of semicarbazide hydrochloride above and below the ferroelectric phase transition. J. Chem. Soc. Faraday Trans. II *72*, 313-323 (1976)
497 P. Figuiere, M. Ghelfenstein. H. Szwarc: First-order phase diagram of thiourea and Raman spectroscopy. Chem. Phys. Lett. *33*, 99-103 (1975)
498 I.J. Fritz: Ultrasonic attenuation and mechanism for the 205 K transition in $BaMnF_4$. Phys. Rev. Lett. *35*, 1511-1514 (1975)
499 I.N. Goncharuk, E.V. Chisler: Vibration of 'inverted' NO_2^- radicals in the Raman scattering spectrum of sodium nitrite. Sov. Phys.-Solid State *18*, 329-331 (1976)
500 W.G. Harter, S.P.S. Porto: "Raman Spectra of Solids Involved in Order-Disorder Phase Transitions", in *Proc. of the Esfahan Symp. on Fundamental and Applied Physics*, Esfahan, Iran, 29 Aug - 5 Sept 1971, ed. by M.S. Feld, A. Javan, N.A. Kurnit (Wiley-Interscience, New York 1973) pp. 629-633
501 J.A. Holy, M.V. Klein, W.L. McMillan, S.F. Meyer: Raman-active lattice vibrations of the commensurate superlattice in 2H-$TaSe_2$. Phys. Rev. Lett. *37*, 1145-1148 (1976)
502 J.A. Holy, K.C. Woo, M.V. Klein: Raman and infrared studies of superlattice formation in $TiSe_2$. Phys. Rev. B*16*, 3628-3637 (1977)
503 E.A. Ivanova, E.V. Chisler: Infrared spectroscopic investigation of a ferroelectric transition in an $NaNO_2$ crystal. Sov. Phys.-Solid State *17*, 1919-1924 (1975)

504 C. Konak, J. Matras: Birefringent and electrooptical properties of $(NH_4)_2BeF_4$. Czech. J. Phys. B *B26*, 577-584 (1976)
505 Yu. A. Popkov, V.F. Shabanov, V.V. Eremenko, K.S. Aleksandrov: Raman studies of the ferroelectric phase transition in $(NH_4)_2BeF_4$. Fiz. Nizk. Temp. *1*, 936-944 (1975) (in Russian)
506 R.A.O. Prasad: "Disorder Induced Raman Scattering in $NaNO_2$", in Ref. 2, pp. 877-882
507 J.F. Ryan, J.F. Scott: Raman study of soft zone boundary phonons and the antiferrodistortive phase transition in $BaMnF_4$. Solid State Commun. *14*, 5-9 (1974);
— J.F. Ryan, J.F. Muratore, J.F. Scott: Light scattering from soft modes in $BaM^{2+}F_4$: a new class of antiferroelectric. Ferroelectrics *7*, 279-281 (1974)
508 J.F. Ryan, J.F. Scott: "Light Scattering Study of Antiferroelectric $BaMnF_4$", in Ref. 2, pp. 761-769
509 S.M. Shapiro, D.E. Cox, R.A. Cowley, M. Eibschutz, H.J. Guggenheim: Soft mode behavior in $BaMnF_4$. Bull. Am. Phys. Soc. *23*, 314 (1978)
510 E.F. Steigmeier, G. Harbeke, H. Auderset, F.J. Di-Salvo: Softening of charge density wave excitations at the superstructure transition in $2H-TaSe_2$. Solid State Commun. *20*, 667-677 (1976)
511 J.C. Tsang, J.E. Smith, M.W. Shafer: Raman spectroscopy of soft modes at the charge-density-wave phase transition in $2H-NbSe_2$. Phys. Rev. Lett. *37*, 1407-1410 (1976)
512 M. Wada, A. Sawada, Y. Ishibashi: Raman spectrum of $(NH_4)_2BeF_4$. J. Phys. Soc. Jpn. *43*, 950-953 (1977)
513 M. Wada, A. Sawada, Y. Ishibashi, Y. Takagi: Raman scattering spectra of K_2SeO_4. J. Phys. Soc. Jpn. *42*, 1229-1234 (1977)
514 M. Wada, H. Uwe, A. Sawada, Y. Ishibashi, Y. Takagi, T. Sakudo: The lower frequency soft-mode in the ferroelectric phase of K_2SeO_4. J. Phys. Soc. Jpn. *43*, 544-547 (1977)

14 *Incommensurate Transitions: Other Optical Techniques*

515 H. Vogt: Study of long and short range order in ferroelectrics by optical second harmonic generation. Ferroelectrics *7*, 103-104 (1974)

15 *Uncategorized References*

516 G. Ahlers, A. Kornblit, H.J. Guggenheim: Logarithmic corrections to the Landau specific heat near the Curie temperature of the dipolar Ising ferromagnet $LiTbF_4$. Phys. Rev. Lett. *34*, 1656-1659 (1974)
517 M. Barmatz, L.R. Testardi, F.J. DiSalvo: Elasticity behavior of $2H-TaSe_2$. Phys. Rev. B*13*, 4637-4639 (1976)
518 P. Bastie, M. Vallade, C. Vettier, C.M.E. Zeyen: Study of tricritical point in KDP by γ-ray and neutron diffractometry. Phys. Rev. Lett. *40*, 337-340 (1978)
519 L.E. Cross, A. Fouskova, S.E. Cummins: Gadolynium molybdate: a new type of ferroelectric crystal. Phys. Rev. Lett. *21*, 812-814 (1968)
520 F.J. DiSalvo: "Experimental Studies of Charge Density Waves", in Ref. 31, pp. 107-137
521 B. Dorner, J.D. Axe, G. Shirane: Neutron scattering study of the ferroelectric phase transition in $Tb_2(MoO_4)_3$. Phys. Rev. B *6*, 1950-1963 (1972)
522 I.J. Fritz: Ultrasonic velocity measurements near the 250 K phase transition in $BaMnF_4$. Phys. Lett. *51a*, 219-220 (1975)
523 M. Iizumi, J.D. Axe, G. Shirane, K. Shimaoka: Structural phase transitions in K_2SeO_4. Phys. Rev. B*15*, 4392-4411 (1977)
524 D.E. Moncton, J.D. Axe, F.J. DiSalvo: Study of superlattice formation in $2H-NbSe_2$, and $2H-TaSe_2$ by neutron scattering. Phys. Rev. Lett. *34*, 734-737 (1975)
525 T. Riste, E.J. Samuelson, K. Otnes:"Critical Scattering of Neutrons in $SrTiO_3$", in Ref. 32, pp. 395-409

526 V.H. Schmidt: Tricritical point in KH_2PO_4. Phys. Rev. Lett. *37*, 839-842 (1976)
527 S.M. Shapiro, R.A. Cowley, D.E. Cox, M. Erbschutz, H.J. Guggenheim: "Structural and Magnetic Transitions in $BaMnF_4$", in *Proc. of the Conference O Neutron Scattering, Galtinburg, Tenn.*, ed. by R.M. Moon (National Technical Information Service, Springfield, Virginia (1976) p. 399)

3. Investigation of Structural Phase Transformations by Inelastic Neutron Scattering

B. Dorner

With 22 Figures

Inelastic scattering of neutrons, as compared to X-ray scattering, is a young technique which has been practiced for a little more than 20 years. Phenomena associated with phase transformations were studied with X-rays long before thermal neutron sources of sufficient flux for inelastic neutron scattering became available. Superstructures related to multiples of the original unit cell and pretransitional effects such as diffuse scattering around possible superstructure Bragg points have been observed. Similarly, phase transitions were studied where the volume of the unit cell remains unchanged, as is the case for proper ferroelectrics.

However, the energy transfers related to the elementary excitations of condensed matter, to the dynamics of phase transitions and critical phenomena could not be resolved because the change of the energy of the X-ray photons is only $\sim 10^{-6}$ of the original photon energy.

An important breakthrough came with inelastic scattering of thermal neutrons as they have wavelengths comparable to X-rays and energies comparable to those of lattice vibrations. Investigations of dynamical critical phenomena are presented in Sects.3.1 and 3.2.

Recently the technique of inelastic neutron scattering has been extended to very high energy resolution (< 0.1 μeV). This high resolution (Sect.3.3.2) allowed the separate study of the various types of molecular motions in molecular crystals such as librations, rotational reorientations, translations, soft modes, and relaxation of clusters. The effects of phase transformations on these motions is discussed in Sect.3.3.3.

A few years ago it seemed that the basic work on phase transformations was completed, as studies were being made on the predictions of dynamic scaling theory, multicritical points, and such "sophisticated" details. But then, in conjunction with low-dimensional conductors, incommensurably modulated structures which were observed long ago by X-ray scattering were rediscovered. The dynamics of such materials has recently attracted considerable interest; see Sect.3.4. Theoretical studies predicted that special excitations, the phase and amplitude modes, should be associated with incommensurable phase transitions. Further, the nonlinear "soliton" excitations which had been known in other branches of physics for many years began to excite interest in the field of phase transitions. Thus, "phasons", "amplitudons", and "solitons" were added to the list of more conventional excitations (phonons, magnons, excitons, etc.) which can be studied by inelastic neutron scattering.

3.1 Inelastic Neutron Scattering

A structural phase transformation manifests itself by a change in the long-range correlations of the atoms in a crystal. These long-range correlations are static with an infinite lifetime. Short-range correlations appear at temperatures near to the phase transformation. They are essentially dynamic with a low frequency or a finite relaxation time τ. To study both types of correlations in space and time, neutrons offer extraordinary possibilities.

The wavelength of a thermal neutron (6 THz $\approx$ 25 meV) with λ = 1.8 Å is comparable to atomic distances in solids, and thus neutron scattering is a very suitable tool to study correlations in space.

At the same time thermal neutrons have energies comparable to excitations in solids, so that inelastic scattering processes yield information about the inner dynamics of a crystal. They allow a detailed study of collective excitations such as phonons and soft modes and of relaxation phenomena with a lifetime τ such as fluctuations of ordered clusters and overdamped soft modes.

With the implementation of cold sources (liquid hydrogen or deuterium) the flux of cold neutrons was enhanced by an order of magnitude. Special high-resolution instruments (discussed in Sect.3.3.1) working with cold neutrons allow one to resolve lifetimes as long as 10^{-7} s. For a comparison, usual lattice vibrations are oscillations whose period is of the order of 10^{-12} s.

3.1.1 General Aspects

The neutrons are characterized by their wave vector $\underline{k}$ in the direction of the beam,

$$\hbar|\underline{k}| = \hbar \frac{2\pi}{\lambda} = m|\underline{v}| \quad , \tag{3.1}$$

where m is the mass of the neutron and v its velocity. The momentum transfer in the scattering process is

$$\hbar\underline{Q} = \hbar\underline{k}_I - \hbar\underline{k}_F \tag{3.2}$$

with $\underline{k}_I$ and $\underline{k}_F$ being the neutron wave vector before and after scattering.

$\hbar = h/2\pi$ (normalized Planck's constant)

The change in energy is

$$\hbar\omega = \frac{\hbar^2}{2m}(k_I^2 - k_F^2) \quad . \tag{3.3}$$

The relation between different energy and frequency units is 1 THz = 4.138 meV = 33.37 cm^{-1}. In an actual experiment the measured scattered intensity is a folding of the scattering function $S(\underline{Q},\omega)$ with the resolution or transmission function R of the spectrometer used. In the following we shall discuss exclusively the scattering function $S(\underline{Q},\omega)$ as it describes the physical phenomena of the substances under investigation. Experimental techniques and resolution problems are extensively discuss- elsewhere, for example, [3.1]. Another review on neutron scattering at structural phase transformations appeared recently by CURRAT and PYNN [3.2]

There one finds an extensive account of the impact of inelastic neutron scattering on the investigation of phase transformations. Therefore this chapter will be short and complementary. We shall mainly discuss phenomena which have not been considered by DORNER and COMES [3.1] because the situation was not clear enough at the time, as for the "central peak" phenomenon. A section will be devoted to high-resolution inelastic neutron scattering and its application in the investigation of phase transformations in molecular crystals. An introduction to the new and highly interesting field of incommensurable structures will be given at the end.

3.1.2 Structural Changes with Small Displacements

The scattering function $S(\underline{Q},\omega)$ of the atoms is the Fourier transform of their position correlations in space and time.

The scattering function integrated over energy $S(\underline{Q})$ (taking the correlations at time equal zero) at a given momentum transfer $\underline{Q}$ (which defines a certain position in reciprocal space) is for periodic structures

$$S(\underline{Q}) \approx \Big| \sum_{d}^{\text{unit cell}} \bar{b}_d \exp[-W_d(\underline{Q}) + i\underline{Q}\cdot(\underline{d} + \underline{u}_d)]\Big|^2 \quad , \tag{3.4}$$

where $\underline{d}$ is the position vector to the equilibrium or high-symmetry position of atom d in the unit cell and $\bar{b}_d$ is the coherent scattering length of atom d. The incoherent scattering is considered as a contribution to the background and therefore neglected for the moment. $W_d(\underline{Q})$ is the exponent of the Debye-Waller factor for atom d and $\underline{u}_d$ is its displacement from the high-symmetry position. Equation (3.4) holds for small and large $\underline{u}_d$'s. The $\underline{u}_d$'s may be static (superstructure) or time dependent (oscillations or relaxation).

For small $\underline{Q} \cdot \underline{u}_d$ we can expand (3.4),

$$\exp(i\underline{Q} \cdot \underline{u}_d) = 1 + i\underline{Q} \cdot \underline{u}_d + \dots \quad . \tag{3.5}$$

The zeroth order of the expansion gives the Bragg scattering of the high-symmetry structure and will not be discussed any further. We focus attention on the first-order term.

If we allow $\underline{u}_d$ to be time and temperature dependent then the scattering function $S(\underline{Q},\omega)$ will contain a function F accounting for the energy and temperature dependence. Different modes j with different energies will contribute to $S(\underline{Q},\omega)$. In general (quasiharmonic approximation) there are 3 n modes for a given phonon wave vector $\underline{q}$ (given position $\underline{Q}$), where n is the number of atoms per unit cell. In connection with a phase transformation it is usually only one mode j which is of particular interest. We split the scattering function $S_j(\underline{Q},\omega)$ into two components: a $\underline{Q}$-dependent structure factor G_j and a spectral function F_j depending on energy $\hbar\omega$ and temperature T,

$$S_j(\underline{Q},\omega) = |G_j(\underline{q},\underline{Q})|^2 \cdot F_j(\omega,\omega_j(\underline{q}),T) \quad . \tag{3.6}$$

The structure factor

$$G_j(\underline{q},\underline{Q}) = \sum_{d}^{\text{unit cell}} \bar{b}_d[\underline{Q} \cdot \underline{\sigma}_d^j(\underline{q})]M_d^{-\frac{1}{2}} \exp[-W_d(\underline{Q}) + i\underline{Q} \cdot \underline{d}] \tag{3.7}$$

contains the eigenvector σ^j of the mode j, which is normalized to unity and describes the pattern of displacements in one unit cell. It has 3 n components. M_d is the mass of atom d. The displacements $\underline{u}_{d,\ell}^j$ are given by

$$\underline{u}_{d,\ell}^j(\underline{q}) = A_d^j \sigma_d^j(\underline{q}) \exp(i\underline{q} \cdot \underline{\ell}) \quad , \tag{3.8}$$

where A^j is an amplitude factor, which may depend on time, temperature and eventually on the position of the unit cell in the crystal, and $\underline{\ell}$ is the vector to the ℓth unit cell. The effect of time and temperature dependence is included in the spectral function F_j. For the dependence on position as it appears for relaxing clusters see DORNER and COMES [Ref.3.1, Sect.3.1.3].

F_j is basically the imaginary part of the dynamic phonon susceptibility.

$$F_j = \frac{1}{1-\exp(-\hbar\omega/kT)} \operatorname{Im}\{(\Omega_j^2 + \Pi_j - \omega^2)^{-1}\} \quad . \tag{3.9}$$

Ω_j is the harmonic frequency of phonon mode j. For a damped harmonic oscillator with the self-energy

$$\Pi_j(\omega,T) = \Delta_j(T) - i\omega\Gamma_j(T) \qquad (3.10)$$

we find

$$F_j = \frac{\omega}{1-\exp(-\hbar\omega/kT)} \frac{\Gamma_j(T)}{[\omega_j^2(T)-\omega^2]^2+\omega^2\Gamma_j^2(T)} , \qquad (3.11)$$

where $\omega_j = |\Omega_j^2 + \Delta_j(T)|^{\frac{1}{2}}$ is the renormalized quasiharmonic frequency. The static mode susceptibility χ is proportional to ω_j^{-2}, which in turn is proportional to the integrated intensity of the mode jif $kT \gg \hbar\omega$.

So far, we considered only coherent scattering to observe collective phenomena. The information which one can get from incoherent scattering will be discussed in Sect.3.3.1 in the context of high-resolution neutron scattering.

3.1.3 Soft Modes

If a crystal undergoes a displacive phase transformation from one ordered structure to another ordered structure, then there exists an order parameter η which is zero in one phase (of higher symmetry) and nonzero in the low-symmetry phase. η describes a pattern of static displacements $\underline{u}_{d,\ell}$ from the equilibrium positions of the ions. This pattern can be described by the eigenvector of a mode j close to the transition temperature T_c; see (3.8). Going away from T_c, the eigenvector of the static displacements may change due to a coupling of the order parameter to other distortions, as described by elastic strain or other modes, for details see [3.1] and articles by DVORAK [3.3] and by THOMAS and HOECK [3.4] in the corresponding theory volume.

We restrict our discussion of the particular mode j which has the same eigenvector $\underline{\sigma}_d^j(\underline{q})$ as the order parameter, to the high-symmetry phase ($\eta = 0$). It is plausible that the restoring forces for this mode will decrease on approaching the phase transformation, and consequently the frequency of this mode will decrease and get soft. The mode frequency approaches zero for a second-order phase transformation as $T \to T_c$ (sometimes it levels off at a finite frequency; see Sect.3.2). It is called the soft mode. Historically the concept of a soft mode was introduced from the temperature dependence of the dielectric constant at paraferroelectric phase transformations by COCHRAN [2.5] and independently by ANDERSON [2.6].

For the investigation of these soft modes in the high-symmetry phase the inelastic scattering of neutrons is still a unique technique. For ferrodistortive transformations the soft mode (at the center of the Brillouin zone) is very often Raman and infrared inactive. For antiferrodistortive transformations the soft mode appears at a boundary of the Brillouin zone, and is therefore accessible by inelastic neutron scattering only.

Soft modes have been observed in many substances and described by (3.11). As long as the mode is not overdamped ($2\,\omega_j^2 > \Gamma_j^2$) the scattered intensity exhibits two maxima, one for energy gain and one for energy loss. In the overdamped case ($2\omega_j^2 < \Gamma_j^2$) there is only one maximum at $\omega = 0$. The transition from underdamped to overdamped soft modes at structural phase transformations is always found to be caused by the decrease of ω_j. Within experimental errors, Γ_j has always been found to be temperature independent.

As an example we present some early results on $LaAlO_3$ [3.7]. The soft mode was observed at the R point (½, ½, ½) of the cubic phase; see Fig. 3.1. For this sample $T_c = 535 \pm 5^\circ$ C was found. Figure 3.2 shows some constant Q scans at the R point at different temperatures. The curves represent a fit of (3.11) to the data. The arrows indicate the quasiharmonic frequency ω_j at each temperature.

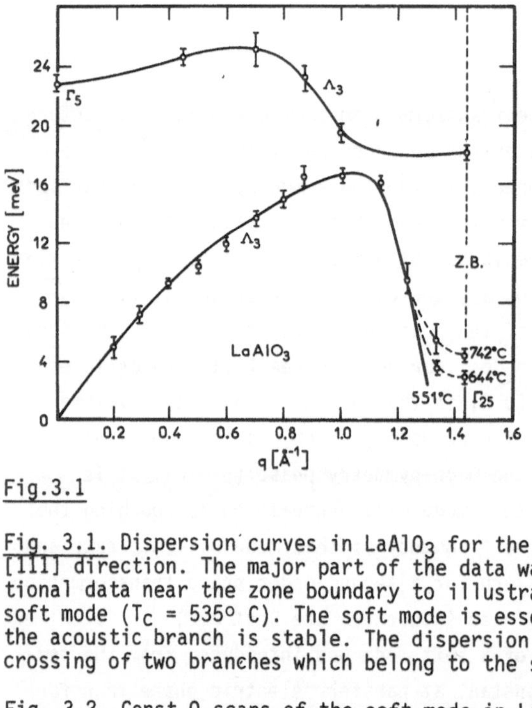

Fig.3.1

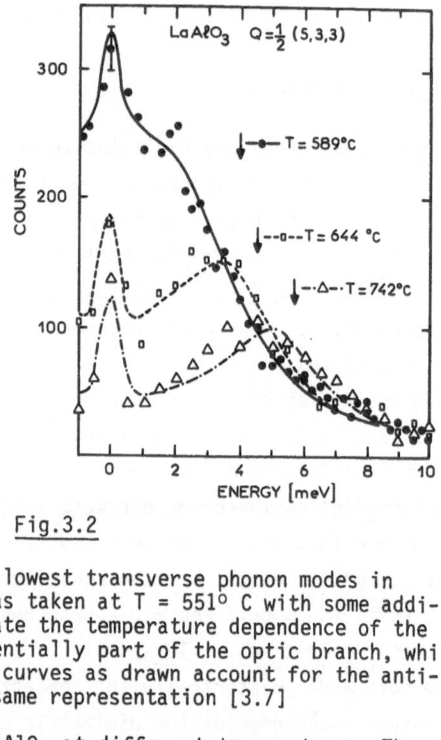

Fig.3.2

Fig. 3.1. Dispersion curves in $LaAlO_3$ for the lowest transverse phonon modes in [111] direction. The major part of the data was taken at T = 551° C with some additional data near the zone boundary to illustrate the temperature dependence of the soft mode (T_C = 535° C). The soft mode is essentially part of the optic branch, while the acoustic branch is stable. The dispersion curves as drawn account for the anticrossing of two branches which belong to the same representation [3.7]

Fig. 3.2. Const Q scans of the soft mode in $LaAlO_3$ at different temperatures. The curves represent least squares fitting of (3.11) with a "background" around $\omega = 0$. The arrows indicate the ω_j at each temperature [3.7]

At 742° C and at 644° C the soft mode is strongly damped, but exhibits a clear maximum in the intensity for energy loss. At 589° C the soft mode is already overdamped. The narrow intensity distribution around $\omega = 0$ was interpreted as general background including incoherent scattering from the sample. The "central peak" intensity (Sect.3.2) was hidden and therefore overlooked for reasons given in Sect. 3.2. After the discovery of the central peak in $SrTiO_3$ [3.8] $LaAlO_3$ was reexamined [3.9] and a central peak could be observed.

3.2 The Central-Peak Phenomenon

The "central peak" was discovered by RISTE et al. [3.8] near the displacive phase transformation in $SrTiO_3$. At the time the phenomenon of a "soft mode" was well known. The striking feature of the central peak is that an intensity appears at $\omega = 0$ while the two soft mode maxima are very well separated; see Fig.3.3. Recently an extended review on the central peak phenomenon has been published by MUELLER [3.10] discussing various experimental techniques as well as different theoretical approaches.

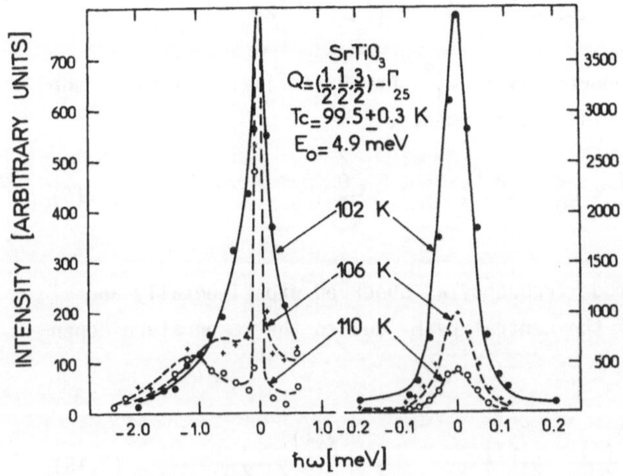

Fig. 3.3. Scattered neutron spectra of $SrTiO_3$ at several temperatures above T_c [3.12]. The left-hand side shows the soft mode behavior of the phonon; the right-hand side shows the divergence of the central peak

The existence of the central peak in $SrTiO_3$ at (1/2, 1/2, 3/2) was proved by RISTE et al. [3.8] because they used the time-of-flight technique which separates the higher order contributions ($2k_I$, $3k_I$, ...). Other groups had overlooked the effect because it was hidden under the intensity coming from $2k_I$ being Bragg scattered from (113). Afterwards, the use of monochromators and analyzers which do not reflect in second order as Ge (111), and the use of good pyrolytic graphite filters [3.11] allowed the investigation of the central peak with three-axis spectrometers [3.12]. The experimental results show that at low frewquencies there is a response additional to that which is described by the "soft mode" formula (3.11), where the damping is proportional to ω. To describe this additional response the self-energy expression can be extended by assuming that the "soft mode" decays into some other mode or combinations of modes and that this further mode decays exponentially with a long characteristic time $1/\gamma$ [3.13]. The extended self-energy can be written as

$$\Pi_j(\omega,T) = \Delta_j(T) - i\omega\Gamma_j(T) - i\omega\frac{\delta^2(T)}{\gamma - i\omega} \quad , \tag{3.12}$$

where $\Delta_j(T)$ and $\Gamma_j(T)$ are the usual renormalization of the frequency and the usual damping constant accounting for ordinary phonon-phonon scattering as in (3.10). The third term describes the coupling (proportional to the coupling constant δ^2) to the mode (still not specified in these days) introduced above. The resulting spectral function F_j of the scattering function reads

$$F_j(\omega,T) = \frac{\omega}{1-\exp(-\hbar\omega/kT)} \frac{\Gamma_j(T) + \dfrac{\gamma\delta^2(T)}{\omega^2+\gamma^2}}{\left[\omega_j^2(T) + \dfrac{\omega^2\delta^2(T)}{\omega^2+\gamma^2} - \omega^2\right]^2 + \omega^2\left[\Gamma_j(T) + \dfrac{\gamma\delta^2(T)}{\omega^2+\gamma^2}\right]^2} \quad . \tag{3.13}$$

The static susceptibility χ is proportional to ω_j^{-2}. For historical reasons of the central peak phenomenon we use the following symbols:

$$\omega_0^2 = \chi^{-1} = \omega_j^2 = \Omega_j^2 + \Delta_j \quad ; \quad \omega_\infty^2 = \omega_0^2 + \delta^2 \quad ; \quad \gamma' = \frac{\omega_0^2}{\omega_\infty^2}\gamma \quad . \tag{3.14}$$

Here ω_∞ is the renormalized soft mode frequency as observed experimentally and γ' the temperature-dependent width of the central peak (due to the temperature dependence of ω_0) see (3.15).

$$F_j(\omega,T) = \frac{\omega}{1-\exp(-\hbar\omega/kT)} \frac{\delta^2(T)}{\omega_0^2(T)\omega_\infty^2(T)} \frac{\gamma'(T)}{\omega^2+\gamma'^2(T)} + \frac{\Gamma_j(T)}{[\omega_\infty^2(T)-\omega^2]^2+\omega^2\Gamma_j^2(T)} \tag{3.15}$$

which could be derived under the assumptions

$$\Gamma_j \ll \delta^2/\gamma \quad ; \quad \omega_\infty^2 \gg \gamma^2 \quad .$$

Equation (3.15) clearly distinguishes two contributions, one Lorentzian with width γ' around $\omega = 0$ and a well-known damped harmonic oscillator (soft mode) expression like (3.11). There are two relaxation times: one proportional to $1/\Gamma_j$ and one proportional to $1/\gamma'$.

Note that SHAPIRO et al. [3.12] started with a different coupling in the self-energy, but that they arrived at the same expression as (3.15). The only difference is that from their Ansatz of the self-energy it follows that $\Omega_j^2 + \Delta_j = \omega_\infty^2$.

The width γ' has not yet been determined by inelastic neutron scattering. Therefore the question is open to what extent the central peak as observed in neutron scattering is dynamic in origin. As in $SrTiO_3$ above T_c precursers of the new structure (soft mode and central peak) appear at the boundary of the Brillouin zone at the R point, inelastic neutron scattering is the only technique to study these phenomena. From high-resolution measurements [3.14] it was found that $\gamma' < 20$ MHz.

ω_∞ and ω_0 are both strongly temperature dependent. ω_∞^2 decreases linearly with temperature for large $T - T_c$ but levels off near T_c and remains finite $\omega_\infty = \delta$ at T_c; see Fig.3.4.

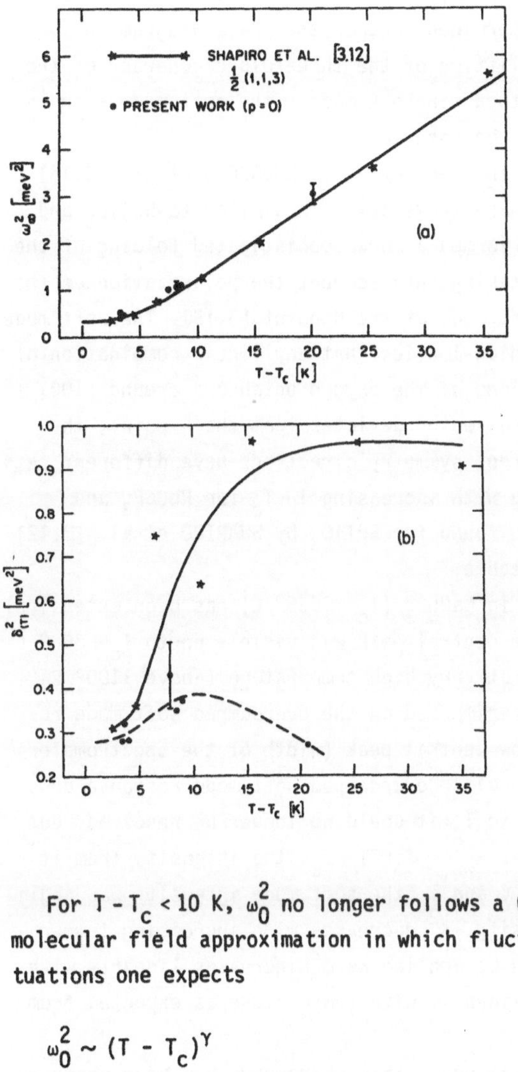

Fig. 3.4a,b. R_{25} soft mode in $SrTiO_3$ in the absence of applied stress. (a) Comparison between two determinations of the quasiharmonic frequency $\omega_\infty(T)$; (b) Some comparison in terms of $\delta^2(T)$, the parameter which characterizes the central peak strength; see (3.15) [3.15]

For $T - T_c < 10$ K, ω_0^2 no longer follows a Curie-Weiss law which is valid in the molecular field approximation in which fluctuations are neglected. Including fluctuations one expects

$$\omega_0^2 \sim (T - T_c)^\gamma \qquad (3.16)$$

with the critical exponent γ. SHAPIRO et al. [3.12] found it very difficult to derive an exponent γ from their data because T_c was uncertain to ± 0.5 K and the quantities ω_0^2, ω_∞^2, δ^2 were extracted from a convolution of the scattering function with the spectrometer resolution. The convoluted function then was fitted to the data. They found δ^2 constant for $T > T_c + 20$ K and decreasing when approaching T_c. The critical exponent γ in $SrTiO_3$ is expected to be the Ising value $\gamma = 1.25$ [3.16] due to the presence of internal strain. For crystals compressed only in [110] direction this value would be correct [3.17].

An investigation of $SrTiO_3$ under [111] uniaxial stress [3.15] confirmed the temperature dependence of ω_∞^2 as obtained by SHAPIRO et al. [3.12], but found an almost constant δ^2 of a value which corresponds to the δ^2 of SHAPIRO et al. [3.12] near T_c; see Fig.3.4. CURRAT et al. [3.15] determined further the phase diagram in T - P ([111] uniaxial) space and verified the lifting of the threefold degeneracy of the soft mode under uniaxial [111] stress into a single A mode - the soft mode - and a doubly degenerate E mode, which does not condense out.

In $LaAlO_3$ KJEMS et al. [3.9] found δ^2 independent of T. ROUSSEAU et al. [3.18] investigated $RbCaF_3$, which undergoes a phase transformation similar to $SrTiO_3$ and exhibits a central peak as well. They performed a more sophisticated folding of the scattering function with the resolution taking into account the polarization of the dispersion branches emerging from the soft mode at the R point [3.19]. The soft mode at the R point is threefold degenerate which implies that any linear combination of the three orthogonal eigenvectors (rotations of the oxygen octahedra around [100] axis) is valid. As soon as the phonon wave vector deviates from the R point, the threefold degeneracy is lifted and different symmetry directions have different sets of eigenvectors. They found δ^2 decreasing with increasing $T - T_c$ for $RbCaF_3$ and suspected that the temperature dependence found for $SrTiO_3$ by SHAPIRO et al. [3.12] might be an artifact of the folding procedure.

The cubic-tetragonal phase transition in $NaNbO_3$ (T_c = 635° C) has been extensively studied by DENOYER and CURRAT [3.20]. The central peak was visible up to T_c + 360° C. As the soft mode became underdamped only at very high temperature (above 1100° C) the central peak intensity was always superimposed on the overdamped soft mode response. By a lineshape analysis the narrow central peak (width of the spectrometer resolution) could be separated from the broader overdamped soft mode response down to about 800° C. For temperatures nearer to T_c it could no longer be resolved. But because the soft mode frequency stabilizes $\omega_\infty \to \delta$ for $T \to T_c$, the intensity from it stabilizes as well. Finally between 680° C and T_c the soft mode intensity was negligible compared to the central peak intensity. δ^2 decreases with increasing temperature. For T > 840° C mean field theory can be applied as ω_∞^2 increases linearly with temperature. In this range δ^2 decreases linearly with temperature as expected from the theory by HALPERN and VARMA [3.21].

Since the first discovery of the central peak, this phenomenon has been observed in many substances and up till now all experimental results could be well described by (3.15). Nevertheless there was no convincing microsopic explanation for the low-frequency decay channel having the long relaxation time $1/\gamma'$.

AXE et al. [3.22] brought up the idea that defects might be at the origin of the central peak. But this model had the defect that the intensity came out to be proportional to ω_0^{-4}. HALPERN and VARMA [3.21] published a theory discussing the possible coupling of defects to lattice vibrations. To allow a linear coupling the strain field around the defect has to have the same group theoretical representation as the soft mode. The intensity of the central peak was proportional to $(\omega_\infty \cdot \omega_0)^{-2}$ as observed experimentally. As it is extremely difficult to characterize a defect and its strain field, an experimental proof of this theory is still lacking. Nevertheless recent observations provide experimental evidence for the involvement of a defect mechanism in the central peak formation.

HASTINGS et al. [3.23] found a central peak enhancement in hydrogen-reduced $SrTiO_3$ with increasing defect concentration. The phase transformation temperature T_c decreases linearly with increasing carrier concentration from 105 K for zero concentration to 70 K for $3.2 \cdot 10^{20}$ carriers per cm^3. The soft mode frequency stays the same for a given $T - T_c$ whatever T_c is. The intensity of the central peak increases considerably less than linearly with the carrier concentration.

Note that the carrier concentration could be determined but not the defect concentration. HASTINGS et al. [3.23] invoked as defect a Ti^{3+} on a Sr^{2+} position displaced towards one oxygen ion. This is a bit surprising as one would expect a lack of oxygen after the reduction with hydrogen [3.24].

Recently WAGNER et al. [3.25] performed further experiments with inelastic neutron scattering on many samples with different degrees of reduction up to a carrier concentration of $2 \cdot 20^{21}$ cm^{-3}. It was observed that for very high carrier concentration the structural phase transformation no longer occurs.

From the theory by HALPERIN and VARMA [3.21] one concludes that the defects in hydrogen reduced $SrTiO_3$ are rather frozen in, because T_c is decreased. The small hopping frequency of the defects is as well reflected in the small γ'. For hopping frequencies larger than ω_∞, T_c is predicted to increase with increasing defect concentration. In this case γ' would be larger than ω_∞ and no central peak should occur.

An interesting case in this context is KCN [3.26], where γ is rather large and its origin is the reorientation of $(CN)^-$ ions. A central peak is only visible for $\omega_j \gtrsim \gamma$. For a detailed discussion see Sect. 3.3.3c and Fig. 3.9.

The central peak phenomenon has also been investigated by other techniques as EPR finding $\gamma' < 6$ MHz [3.27,28] and by Raman scattering [3.29]. Unfortunately a comparison of the results with those from neutron scattering is difficult, because EPR is a local probe in real space and Raman scattering is limited to the origin of reciprocal space. Recently, however, the local probability distribution obtained by EPR, the soft mode frequency below T_c obtained by Raman scattering, and the central peak soft mode spectrum above T_c obtained by inelastic neutron scattering have been related empirically by BRUCE et al. [3.30].

Sometimes the expression "central peak" is used for effects which have nothing to do with the central peak as described above. For example the critical scattering at a disorder-order phase transition, which has only one temperature-dependent relaxation time, should not be called a central peak, although the scattering is centered around $\omega = 0$. A "central peak" would appear if a second much longer relaxation time came into play.

3.3 Molecular Crystals

The centers of gravity of molecules in molecular crystals are usually well ordered often in contrast to the orientations of the molecules. Orientational interactions (as due to multipoles for example) at elevated temperatures are in many cases weak compared to thermal energy. If the symmetry of the molecule is equal to or higher than the site symmetry, orientational order grows continuously with decreasing temperature. Only if the symmetry of the molecule is lower than the site symmetry, is a phase transformation necessary to produce long-range orientational order.

Many molecular crystals exhibit phase transformations from orientational disorder to order and from one order to another as well. The disorder ranges from almost complete disorder in orientation to an equal distribution over a few symmetry preferred orientations. The angular difference between such symmetry orientations is large (e.g., 90° or 120°). That means that the displacements $\underline{u}_d$ of the atoms of the molecule are large. Therefore we cannot anymore expand (3.4). One has to describe the density distribution by spherical harmonics as introduced by PRESS and HÜLLER [3.31]. In the case of coherent scattering this technique is used to describe the interference between different atoms.

A particular case arises if incoherent scattering is the dominating scattering process. The incoherent intensity is the sum of the incoherent intensities of each individual atom. There is no interference between scattering from different atoms. The incoherent scattering function $S_{inc}(\underline{Q},\omega)$ is the Fourier transform of the self-correlation function of one atom in space and time.

3.3.1 The Elastic Incoherent Structure Factor (EISF)

If the atom has a definite equilibrium position, then S_{inc} is a monotonic function of Q. The situation changes dramatically if the atom can diffuse. Let us assume that an atom can hop between two positions, one at $\underline{R} = 0$ and one at $\underline{R} = \underline{a}$. At time $t = 0$ it will be at $\underline{R} = 0$ and then it will go over the $\underline{R} = \underline{a}$ with a jump rate λ. The correlations in space $g(\underline{R})$ and time $f(t)$ are then

$$\sum^{\text{sites}} g(\underline{R})\cdot f(t) = \delta(\underline{R})[\tfrac{1}{2} + \tfrac{1}{2}\exp(-2\lambda t)] + \delta(\underline{R} - \underline{a})[\tfrac{1}{2} - \tfrac{1}{2}\exp(-2\lambda t)] \quad . \tag{3.17}$$

The Fourier transform in $\underline{R}$ yields

$$S(\underline{Q},t) \sim [\tfrac{1}{2} + \tfrac{1}{2}\cos(\underline{Q}\cdot\underline{a})] + [\tfrac{1}{2} - \tfrac{1}{2}\cos(\underline{Q}\cdot\underline{a})]\exp(-2\lambda t) \quad . \tag{3.18}$$

After Fourier transformation in time we find

$$S_{inc}(\underline{Q},\omega) \sim [\tfrac{1}{2} + \tfrac{1}{2}\cos(\underline{Q}\cdot\underline{a})]\delta(\omega) + [\tfrac{1}{2} - \tfrac{1}{2}\cos(\underline{Q}\cdot\underline{a})]\,\frac{1/\pi\cdot 2\lambda}{\omega^2+(2\lambda)^2}$$

$$\sim G_1(\underline{Q},\underline{a})\delta(\omega) + G_2(\underline{Q},\underline{a})\,\frac{1/\pi\cdot 2\lambda}{\omega^2+(2\lambda)^2} \quad . \qquad (3.19)$$

To simplify the expression we left out the Debye-Waller factor and the incoherent cross section. For details see the review by LEADBETTER and LECHNER [3.32]. The δ function in energy appears because for $t\to\infty$ there is a finite probability of finding the particle at $\underline{R} = 0$ and at $\underline{R} = \underline{a}$. The second term is called quasielastic scattering. The G_ν are the structure factors.

To distinguish the elastic from the quasielastic scattering experimentally one needs a resolution in energy $\Delta\omega < 2\lambda$. To study rotational diffusion of molecules one needs sophisticated instruments, which will be discussed in Sect.3.3.2. If the resolution is insufficient, then the experiment integrates over energy for each $\underline{Q}$. Inspecting (3.19), we see immediately that the Q dependence disappears as soon as we integrate over energy.

For the one-dimensional hopping atom model (which can be applied, for example, to the protons in KDP) presented above the residence time τ_{res} is $1/\lambda$. But the width of the quasielastic scattering is proportional to 2λ, which reflects $1/(2\cdot\tau_{res})$. If we modify the model into two dimensions, the atom is hopping between two sites, which are situated on a circle. Thus it has two possible ways to go from one position to the other. If the jump rate is still the same the residence time $\tau_{res} = 1/2\,\lambda$. The quasielastic width is again $2\,\lambda$; thus it reflects directly τ_{res} in the two-dimensional model.

The residence time is generally defined by

$$1/\tau_{res} = \sum_{\ell}^{\text{types of jumps}} n_\ell\lambda_\ell \quad , \qquad (3.20)$$

where ℓ labels different types of jumps as, for example, twofold, threefold, fourfold, etc., rotational jumps. λ_ℓ is the jump rate for a particular type and n_ℓ the multiplicity of this type.

Apparently the width of the quasielastic scattering is not a direct measure of a jump rate nor a residence time. But it is proportional to relaxation times τ_ν of different density distributions, which will be discussed in the following.

The scattering function gets more complicated if the hopping atom is part of a molecule, which performs rotational jumps, for example, of 90° with jump rate λ, such that the atom is distributed over 4 positions r_j; see Fig.3.5. From (3.20) with $\ell = 1$ and $n_\ell = 2$, one obtains the residence time equal to $1/(2\lambda)$.

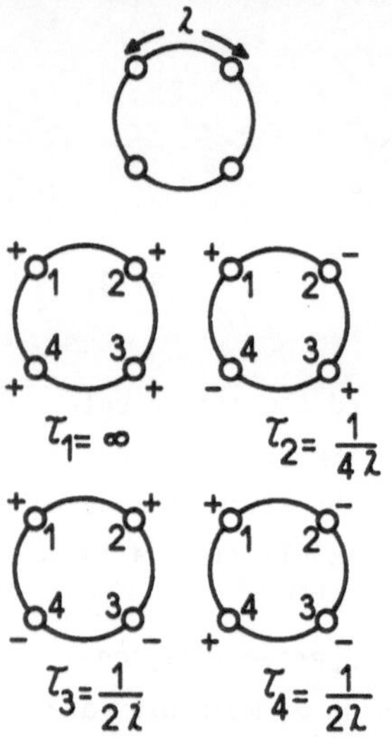

Fig. 3.5. The four basis functions of the density distribution for a particle which can occupy each of four positions with equal probability. The jump rate is λ. The relaxation times for the different density distributions are $\tau_1 - \tau_4$

We write the jump rates in the form of a matrix

$$\bar{M} = \begin{pmatrix} -2\lambda & \lambda & 0 & \lambda \\ \lambda & -2\lambda & \lambda & 0 \\ 0 & \lambda & -2\lambda & \lambda \\ \lambda & 0 & \lambda & -2\lambda \end{pmatrix} . \qquad (3.21)$$

Equation (3.21) means that from position 1 the atom can jump to positions 2 or 4, but not to 3, and that the probability of an atom arriving at position 1 is 2λ.

The eigenvectors σ_ν of this matrix are the basis functions of the density distribution

$$\sigma_1 = \tfrac{1}{2}\begin{pmatrix} 1 \\ 1 \\ 1 \\ 1 \end{pmatrix} \; ; \quad \sigma_2 = \tfrac{1}{2}\begin{pmatrix} 1 \\ -1 \\ 1 \\ -1 \end{pmatrix} \; ; \quad \sigma_3 = \tfrac{1}{2}\begin{pmatrix} 1 \\ 1 \\ -1 \\ -1 \end{pmatrix} \; ; \quad \sigma_4 = \tfrac{1}{2}\begin{pmatrix} 1 \\ -1 \\ -1 \\ 1 \end{pmatrix} . \qquad (3.22)$$

The relaxation times τ_ν for each density distribution we find from

$$\bar{M}\sigma_\nu = -\frac{1}{\tau_\nu}\sigma_\nu$$

to be $\tau_1 = \infty \; ; \quad \tau_2 = \frac{1}{4\lambda} = \frac{1}{2}\tau_{res} \; ; \quad \tau_3 = \tau_4 = \frac{1}{2\lambda} = \tau_{res}$. (3.23)

σ_1 the equal distribution is connected to the δ function in energy. σ_2 has a shorter lifetime than σ_3 and σ_4, because a given position can relax to both sides. A circular random walk model with N sites has been discussed by BARNES [3.33].

The scattering function is a sum of 4 terms, each term being the Fourier transform of $[\sigma_\nu \cdot \exp(-t/\tau_\nu)]$.

$$S_{inc}(\underline{Q},\omega) \sim G_1(\underline{Q},\underline{r}_j)\delta(\omega) + \sum_{\nu=2}^{4} G_\nu(\underline{Q},\underline{r}_j) \frac{1/\pi\ 1/\tau_\nu}{\omega^2+1/\tau_\nu^2} \quad . \tag{3.24}$$

The superposition of Lorentzians with different widths, (3.24), renders the comparison with experimental data problematic. Sometimes one can select $\underline{Q}$ values in lengths and direction (the latter if the sample is a monocrystal) such that one G_ν is dominating, as done by TÖPLER et al. [3.34].

The structure factor G_1 (elastic scattering) is generally called the "elastic incoherent structure factor". This elastic incoherent structure factor contains the information about the preferred axis for the jump rotations as threefold, fourfold, or random. A real system well investigated is NH_4Cl having eight positions, which can be occupied by a particular proton (see Sect.3.3.3).

3.3.2 High-Resolution Instruments

The resolution ΔE_I of a monochromator determines the length in time of the wave packet of the neutron to be $\hbar/\Delta E_I$. If one wants to study the distribution of an incoherent scatterer like a proton, which is distributed over several positions, then the passing wave packet of the neutron has to be long enough to experience one and the same proton at different positions. If the relaxation time is long, the energy resolution has to be very good.

We shall present three different instruments, which are in operation at the Institut Laue Langevin, Grenoble [3.35], connected to the cold source by neutron guides [3.36]. The flux of slow or cold neutrons (0.5 - 1 THz) from a reactor is low because the maximum of the Maxwellian spectrum is at about 10 THz. To increase the flux of cold neutrons, several reactors have a cold source which shifts the spectrum towards lower energies by further moderating the neutrons in liquid hydrogen or deuterium.

a) *The Backscattering Spectrometer*

The Backscattering spectrometer [3.37] works in principle as a three axis spectrometer [Ref.3.1, Sect.3.1.2] with extreme angles at monochromator and analyzer. The energy resolution ΔE_I of a crystal monochromator is given by

$$\Delta E_I = 2E_I[\cot(\theta_M) \cdot \Delta\theta_M + \frac{1}{n}] \quad , \tag{3.25}$$

where θ_M is the Bragg angle of the monochroamtor, $\Delta\theta_M$ the divergence of the beam, and n is the number of lattice planes contributing to the reflection. n is proportional to the penetration depth into the monochromator, limited by primary extinction.

One sees immediately: the smaller the energy E_I the better the resolution. But neutrons with smaller energies have smaller wave vectors and thus the range of momentum transfer $|\underset{\sim}{Q}|$ is restricted.

The $\cot(\theta_M)$ vanishes for $\theta_M = 90^\circ$, that is, backscattering from the monochromator. This leaves us with 1/n to determine the resolution. One uses perfect crystals like Si with polished surfaces and obtains $1/n \approx 4 \times 10^{-5}$. This means 2.5×10^4 atomic planes are the limit for primary extinction. If secondary extinction comes into play due to nonperfect crystals, the resolution deteriorates rapidly.

The instrument IN10 at ILL (a similar instrument is operating at the Dido-Reactor, Jülich [3.38]) achieves a ΔE_I of 80 MHz using backscattering at the monochromator and the analyzer. The range of energy transfer is limited to the speed of the Doppler drive system of the monochromator. For details see [3.35].

b) *The Four Chopper Time-of-Flight Spectrometer*

In the time-of-flight technique the energy of the neutrons is determined by determining their velocity. A first chopper (15,000 r/min) produces a short pulse, which spreads afterwards due to the different velocities of the neutrons. A second chopper several meters behind the first and having a electronically controlled phase with respect to the first, produces a short pulse of neutrons with a certain velocity. The resolution depends on the pulse lengths (speed of the choppers) and the distance between the choppers. The instrument IN5 at the ILL [3.39] has neutron guides between the choppers to avoid intensity losses by the distance. A third chopper suppresses higher order neutrons having smaller velocities than the nominal ones and a fourth chopper regulates the repetition rate of the pulses to avoid an overlap of the intensity distribution at the detectors.

Banks of detectors are installed 4 meters away from the sample. The arrival time of the neutrons depends on the change in their velocity (energy) at the scattering event.

c) *The Neutron Spin-Echo Spectrometer*

The neutron spin-echo technique works with polarized neutrons. After polarization the neutrons pass through a magnetic field with field direction perpendicular to the spin direction, where they perform several thousand Larmor precessions before hitting the sample. After scattering the neutrons pass through a magnetic field of the same strength and length as the first but opposite field (in principle) direction so that the elastically scattered neutrons perform the same number of Larmor precessions in opposite sense as before the sample. At a polarization analyzer all elastically scattered neutrons (whatever their energy is) appear to be polarized as before entering the first magnetic field. In practice the two magnetic fields have the same field direction such that the guide field never drops to zero along the path of the neutron (this would depolarize them). But a π spin turn device near the sample inverts one component of the neutron spin which is perpendicular to the magnetic field. The resolution is proportional to the number of Larmor precessions. As each neutron in this technique is treated individually - precessed forwards before scattering and backwards after scattering - a rather broad band of neutron energies can be tolerated of about 20 to 40% in energy. This broad band provides intensity. The neutron spin-echo spectrometer IN11 [3.40] has been in operation at the ILL since 1978.

3.3.3 Dynamics of Molecular Crystals

Molecular crystals have more atoms per unit cell than many of the other substances investigated so far. This problem is considerably reduced by the assumption that the molecules are rigid bodies bound into the lattice by much weaker forces than the internal ones. This assumption, which turned out to be reasonable in many cases, attributes six degrees of freedom to each molecule; three translational and three rotational.

We shall briefly characterize the phase transformations of some molecular crystals before going into detailed discussion in Sects.3.3.3a-d.

NH_4Cl and ND_4Cl undergo disorder-order (two possible orientations of the ammonia molecule/ferro order) transitions at 242 K and 249 K, respectively. Both substances have at high temperature the same structure as NH_4Br and ND_4Br, where halide ions occupy the corners of a cube and the ammonia is body centered. NH_4Br and ND_4Br undergo phase transformations from the same disorder to an antiferro order at 235 K and 215 K, respectively (the isotope effect is not understood). At lower temperature (hyteresis 80 K - 115 K) NH_4Br and ND_4Br have a strongly first-order phase transformation to ferro order.

Adamantane $C_{10}H_{16}$ and $C_{10}D_{16}$ goes from orientational disorder over two possible orientations to an antiferro order at 208.7 K and 208.3 K, respectively. The structure is fcc above T_c with one molecule per unit cell, and tetragonal below the transition with two molecules per unit cell. The transition is of first order and connected with a volume change of 3.9%. Therefore crystals big enough for inelastic neutron scattering break usually when cooling through the transition.

CD_4 has two phase transformations. In phase I (27 K<T<89.7 K) the molecular orientations are almost completely disordered. The structure is fcc with one molecule per primitive cell. In phase II (22.1 K < T<27 K) three out of four molecules are orientationally ordered, while one is still disordered [3.41]. The structure is still fcc with eight molecules per primitive unit cell. In phase III below 22.1 K all molecules are ordered. The structure is not known. It may be tetragonal with 16 molecules per unit cell [3.42].

CH_4 has only one phase transformation at 20.4 K, which corresponds to the I-II transition in CD_4. The lower phase transformation does not take place. At low temperature, free rotation of the still disordered molecule could be observed [3.43,44].

KCN exhibits two phase transformations. For literature see REHWALD et al. [3.45]. It is of NaCl type structure above 168 K. The $(CN)^-$ ions are orientationally disordered with some preference for the [111] directions. Below 168 K the CN molecules are oriented in [110] directions, but their dipole moment C-N stays disordered. The cell is orthorhombic. The 168 K phase transformation is strongly first order, while the second one at 83 K is continuous. Below 83 K the dipoles order in an antiparallel manner.

a) *Librations*

Figure 3.6 shows the dispersion curves of ND_4Cl at 85 K [3.46,47]. The highest dispersion curve represents the librations of the ND_4 molecule. It is flat for all directions, and one might call the molecule an Einstein oscillator. This means that each molecule librates independently of the neighboring molecules, because the Cl^- lattice separates them. The potential for rotations is shown in Fig.3.7. The values correspond to NH_4Cl. From Raman, infrared, and neutron scattering [3.48] it is known that the librational frequencies in the ammonium halides decrease slightly and broaden with increasing temperature, but do not change at the phase transition. This means that the curvature at the bottom of the potential Fig.3.7 is almost independent of temperature.

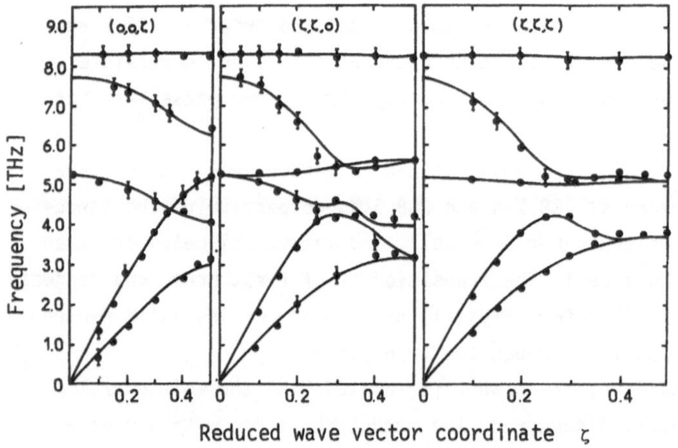

Fig. 3.6. Dispersion Curves for ND_4Cl at 85 K. Experimental points were obtained by TEH and BROCKHOUSE [3.46]; the lines are the best shell model fit by COWLEY [3.47]

Figure 3.8 shows dispersion curves of deuterated adamantane [3.49,50] in the disordered phase above the phase transformation. The librations exhibit some dispersion indicating that the librations are collective motions with a phase relation between neighboring molecules. By means of the scattering geometry transverse and longitudinal librational modes were observed but their frequencies do not differ within experimental uncertainty. The librations are heavily damped probably due to single molecule reorientation as will be discussed in Sect.3.3.3b. Their energy width is of the order of their frequency.

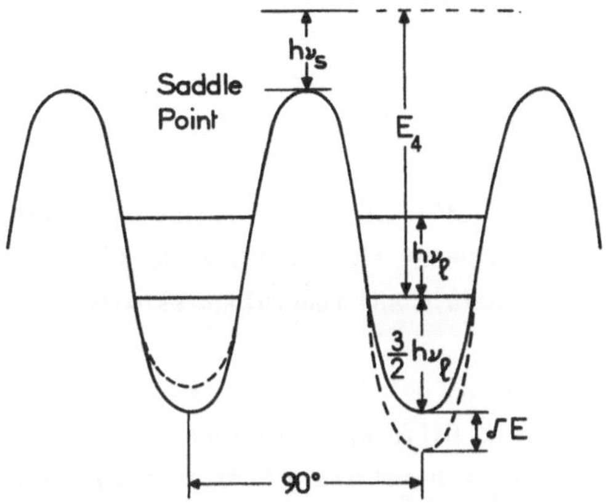

Fig. 3.7. Potential for rotations of a molecule around a fourfold axis. Solid line for the disordered phase with equal distribution; dashed line for the ordered phase with preferred orientation in the right-hand minimum. ν_ℓ is the libration frequency; $h\nu_S$, the ground state on the saddle point; E_4, the activation energy for fourfold jumps; δE, the increase of the activation energy in the ordered phase. Values for NH_4Cl [3.34,55] are: $h\nu_\ell$ = 11.85 THz; $\nu_S = \nu_\ell$; E_4 = 45.5 THz; δE = 5.8 THz, in good agreement between theory and experiment

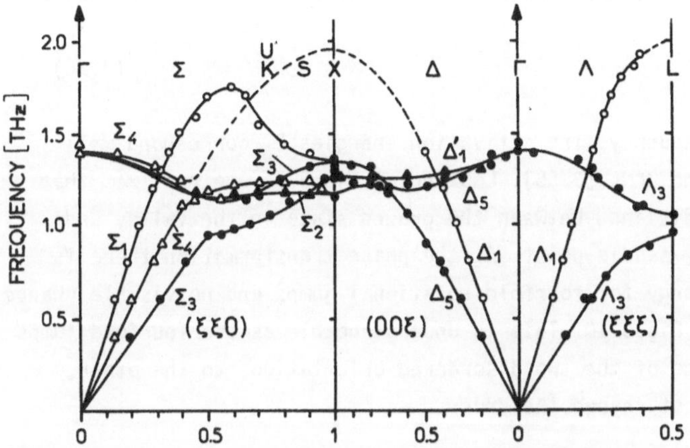

Fig. 3.8. Dispersion curves of adamantane ($C_{10}D_{16}$) at room temperature. Different symbols are used for different representations. As there is one molecule per primitive cell, one expects three acoustic (translational) and three optic (librational) branches. The longitudinal optic mode in the Δ direction was not observed. The high-frequency translational branch around the X point could not be found. Mode anticrossing effects can be seen, where the acoustic and the optic branch belong to the same representation [3.49,50]

In CD_4 no librons could be observed [3.51] either in phase I (disordered or in phase II (partially ordered).

Molecular crystals, which exhibit displacive (order-order) phase transformation form a special group. This means that the molecules go (almost) continuously from one ordered orientation to another ordered orientation. In this case, one expects a soft mode of librational character. Librational soft modes have been observed in chloranil [3.52] above and below the antiferrodistortive phase transformation at 90.3 K. The second example known so far is biphenyl, which is discussed in detail in Sect.3.3.3d.

b) *Rotational Reorientation of a Single Molecule*

As already explained in Sect.3.3.1, incoherent high-resolution inelastic neutron scattering allows one to determine residence time and jump rates λ_ν of reorienting molecules and the preferred axis for these rotational jumps.

Very elaborate work has been performed on NH_4Cl [3.34]. The calculations have been worked out by MICHEL [3.53] and for NH_4Br by LIVINGSTON et al. [3.54]. In NH_4Cl there are reorientational jumps possible around threefold and fourfold axes with λ_3 and λ_4.

To study the two relaxation times as individually as possible TÖPLER et al. [3.34] worked on a single crystal and made utmost use of the different structure factors. For example: $\underline{Q}$ parallel [111] and $|\underline{Q}| = 3.73\ \text{Å}^{-1}$, $G_1 = 0$ and the others such that 78% of the scattered intensity was related to λ_4 and only 22% to a mixture of λ_3 and λ_4; see Fig.3.9. They found for both a temperature dependence following an Arrhenius law

$$\lambda_\nu = \nu_\ell \exp(-E_\nu/kT) \quad , \tag{3.26}$$

where ν_ℓ was the libration frequency. The activation energies E_ν correspond well with calculations by HOLLER and KANE [3.55]. Looking at Fig.3.7, one realizes that the activation energy is the distance between the ground state in the valley and the lowest energy state on the saddle point. At the phase transformation there is a change in the activation energy for fourfold rotational jumps and no visible change for the threefold jumps - see Fig.3.10. This is understandable as the fourfold jumps bring the NH_4 molecule from one of the two disordered orientations to the other, while the threefold jumps do not change the order.

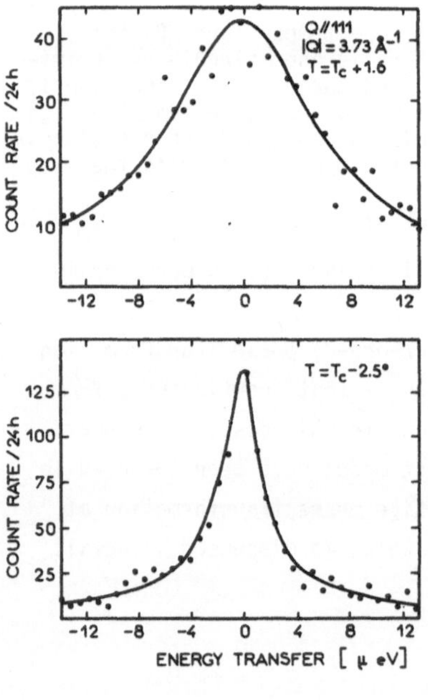

Fig. 3.9. Quasielastic scattering from NH_4Cl above and below the phase transformation $T_c = 242$ K [3.34]. At this Q, elastic scattering is absent (see text). The different energy widths come from different relaxation times and they in turn represent a change in the activation energy for fourfold jump rotation; compare Fig.3.7

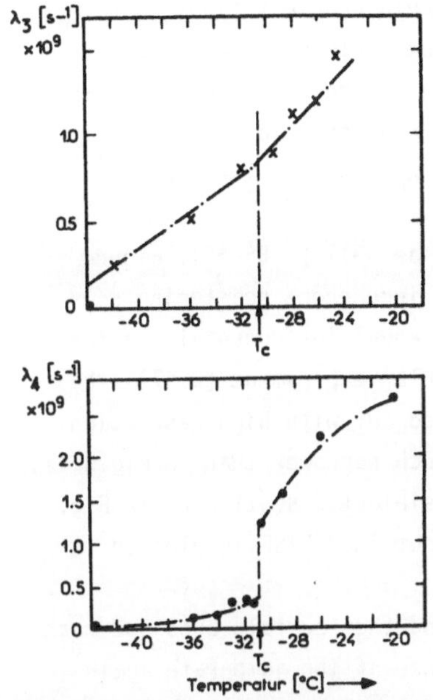

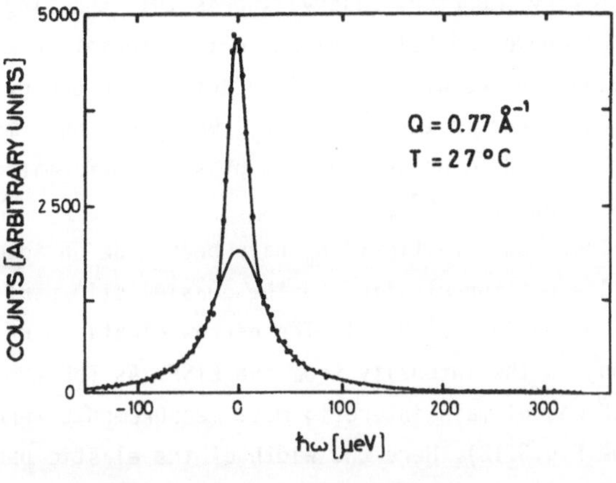

Fig. 3.11. Elastic and quasielastic scattering from adamantane. The lines represent a fit model for fourfold jump rotation. Note the separation into purely elastic and quasielastic parts [3.57]

Fig. 3.10. Jump rates in NH_4Cl in the neighborhood of the phase transformation. λ_3 for threefold jump rotation and λ_4 for fourfold jump rotation [3.34]

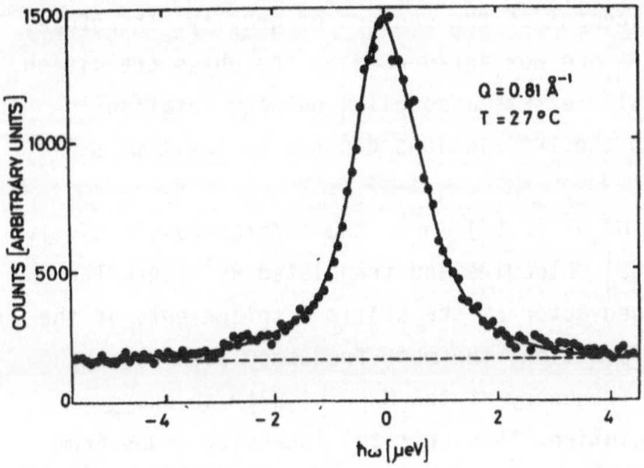

Fig. 3.12. The purely elastic spectrum from Fig.3.11 measured with very good resolution [2.57]. Solid line gives the instrumental resolution and the dashed line represents the quasielastic scattering which is degenerate to a flat background

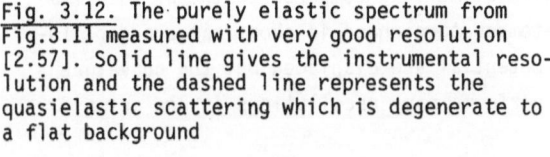

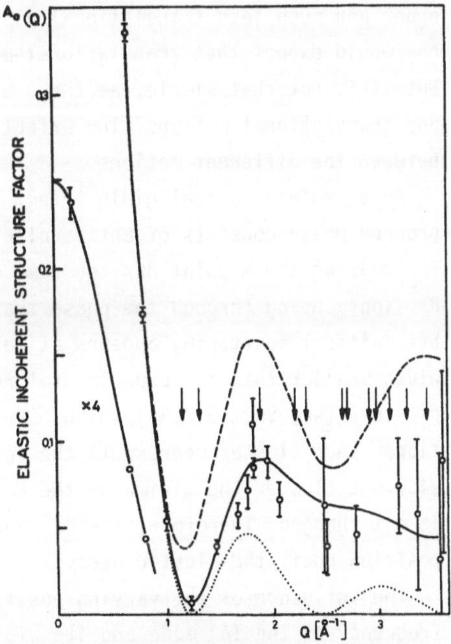

Fig. 3.13. Elastic incoherent structure factors (purely elastic scattering) versus momentum transfer Q for adamantane at room temperature. o are experimental points with statistical error bars; —— model calculation from fourfold jump rotations; --- model calculation from threefold jump rotations; ···· model calculation from diffusion on a sphere; the arrows indicate positions of Bragg peaks [2.57]

A very important observation is that the single molecule reorientations take place above and below the phase transformation with a jump rate governed by an activation energy, (3.26). The activation energy for the disordering fourfold jumps is different in the two phases. But the single molecule reorientation is certainly not related to or driving the phase transformation. The λ_4 above is $1.2 \cdot 10^9$ s^{-1} and below $0.4 \cdot 10^9$ s^{-1}.

Thorough investigations have been made on adamantane as well [3.56-58]. Figure 3.11 shows the elastic and the quasielastic part of the incoherent scattering as discussed in Sect.3.3.1. The narrow elastic part (width equal instrumental resolution) is the intensity from the EISF. As for the particular experiment [2.57], the EISF was of main interest; most measurements were carried out with high resolution (see Fig.3.12). Here the width of the elastic part is much narrower than in Fig.3.11, but still given by the instrumental resolution. The quasielastic scattering in Fig. 3.11 degenerates to a flat background in Fig.3.12. The obtained EISF is plotted in Fig.3.13 and compared with several model calculations.

It was found that the reorientations take place around the fourfold axes and much less around the threefold axes. The temperature dependence of the jump rate again followed an Arrhenius law [3.58].

c) *Translational Lattice Vibrations*

Phase transformations in molecular crystals exhibit mainly a change in orientational order and much less translational displacement of the centers of gravity. Therefore, one would expect that translational modes are not influenced by the phase transition. But it is not that simple, as there usually exists a coupling between rotational and translational motions. The effect on the translations depends on the time scale between the different motions.

An example of a negligible effect is ND_4Br [3.59] where the antiferrodistortively ordered phase consists of antiparallel ND_4^+ molecules and translated Br^- ions. The TA_2 mode at the M point has the same eigenvector as the static displacements of the Br^- ions. Going through the phase transformation, the mode frequency does not change but critical scattering appears at zero frequency at the M point with an energy width smaller than the experimental resolution. This critical intensity comes from clusters (see Sect.3.3.3d) which have a lifetime long compared to the Br^- oscillations. Each cluster represents the low-temperature phase in short range order. Inside one cluster the minima of the Br^- potential are displaced but the curvature is not changed. Therefore, the Br^- ion performs many oscillations in one actual position until the cluster decays.

The influence of the varying position of the minimum of the Br^- potential on the frequency of the TA_2 mode and its width is very small and could not be observed experimentally.

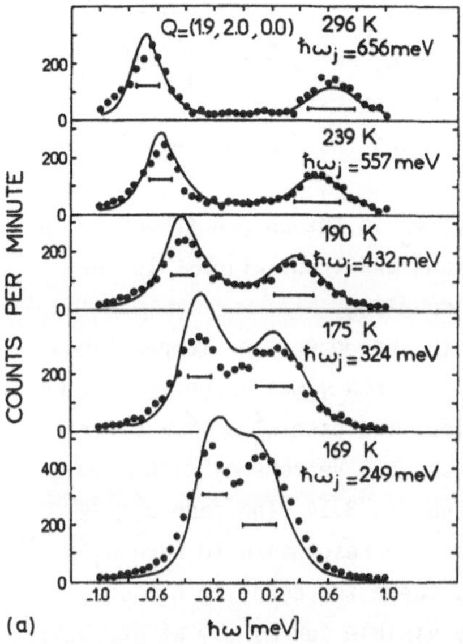

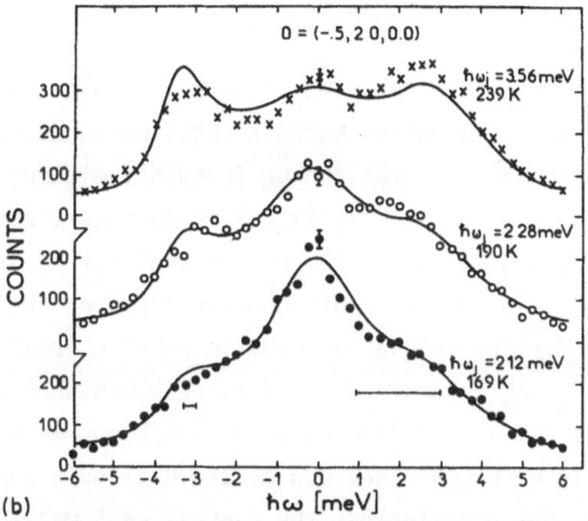

Fig. 3.14. (a) A comparison of neutron lineshapes for KCN for a transverse acoustic mode with wave vector q equal to 1/10 of the Brillouin zone distance in the [100] direction with (3.13). The horizontal bars represent calculated resolution widths for a planar dispersion surface. The only parameter varied is the energy of the mode. The data are not corrected for spectrometer efficiency; rather, the theory is converted to the spectrometer representation. The discrepancies for T = 175 and 169 K are believed to be due to the inadequacy of the resolution correction used. Note that although the mode softens (this mode is related to C_{44} as $q \to 0$), there is no "central peak", i.e., scattering at $\hbar\omega = 0$ [3.26]. (b) Comparison of neutron scattering lineshapes for the same branch as in (a) with (3.13) for a wave vector q = 0.5 (1/2 distance to the Brillouin zone boundary). Note that here $\hbar\omega_j$ is about 3 meV and thus comparable to the orientational relaxation parameter $\hbar\gamma$ = 3 meV. As the temperature is lowered towards 168 K, the behavior is quite different from that in (a). In this figure, there are no spectrometer efficiency corrections as the data were taken with fixed final energy. The peaks broaden and the intensity at $\omega = 0$ increases, confirming the different behavior of the theory and experiment for $\omega_j^2 \ll \gamma^2$ and $\omega_j^2 \sim \gamma^2$

Strong coupling appears in KCN where an effective $(CN)^- - (CN)^-$ ion interaction is mediated by elastic strains. Above the first-order phase transformation at about 168 K the elastic constant C_{44} (related to shear strains along cube faces which tend to distort the cubic cell towards the orthorhombic structure) softens markedly with decreasing temperature [3.60]. C_{44} extrapolates to zero at 154 K, about 14 K below the actual phase transformation.

Inelastic neutron scattering experiments [3.26,61] showed that the TA branch corresponding to C_{44} softens far into the Brillouin zone. A theory by MICHEL et al. [3.62], and MICHEL and NAUDTS [3.63] describing the coupling between reorientation and translations was used to analyze the lineshapes. They arrived at the same expression as (3.13).

The phonon (transverse acoustic in KCN) will appear near to $\omega_j^2 + \delta^2\omega^2/(\omega^2+\gamma^2)$. If $\omega_j^2 >> \gamma^2$, the quasiharmonic phonon frequency is $\omega_\infty^2 = \omega_j^2 + \delta^2$ and if $\Gamma_j << \delta^2/\omega_j$ then (3.15) describes the whole response including the central peak.

If $\omega_j^2 << \gamma^2$, the quasiharmonic frequency is $\omega_0^2 = \omega_j^2$ and the central peak is absent; see also [3.21].

In this case, where ω_0 is a measurable frequency, ω_0 is the phonon frequency including the coupling between rotational reorientations and translations. In the first case ω_∞ is the phonon frequency without this coupling.

For KCN ROWE et al. [3.26] found that γ was nearly independent of temperature and Γ_j negligibly small. But in contrast to $SrTiO_3$, for KCN γ was large ($\hbar\gamma$ = 3 meV), that is in the order of acoustical phonon frequencies. The case $\omega_j^2 \ll \gamma^2$ was verified for low-frequency acoustic phonons (small phonon wave vectors q) with $\omega_j \approx$ 0.25 γ. As expected no central peak was observed; see Fig.3.14. The case $\omega_j^2 \gg \gamma^2$ could not be verified because the phonon wave vectors corresponding to such ω_j's had to be bigger than 0.5 of the Brillouin zone and there the coupling δ^2 goes to zero. But nevertheless the central peak was already visible for $\omega_j \approx \gamma$ at q = 0.5; see Fig.3.14.

Critical scattering was absent because C_{44} goes to zero before the angular correlations become infinite which is as well reflected in the fact that γ did not vary considerably with temperature.

Long-wavelength low-energy acoustic phonons are generally observed in disordered molecular crystals, because the long wavelengths average over the disorder and the small frequencies make a decay into a single molecular reorientation very improbable. In CD_4 neither for phase I nor II could acoustic phonons be observed beyond 2 THz [3.51].

d) *Soft Modes and Relaxation of Clusters*

AUBRY [3.64] made a comparison of the mechanisms for order-disorder and displacive type phase transitions. Figure 3.15 shows possible effective local single-molecule potentials for the different types. The disorder may consist in an equal distribution over two sites (see Fig.3.15a) separated by a potential barrier. The transition from one site to the other is a single-particle relaxation process governed by an activation energy. It is possible that vibrations (in case of translational disorder) or librations (in case of orientational disorder) exist in each well as discussed in Sect.3.3.3a. Below the phase transformation temperature, only one site is occupied in long-range order. The potential of the occupied site is a bit lower than of the nonoccupied site. This difference is usually considered to be small compared to the potential barrier between the sites. Near to the phase transformation critical quasielastic scattering appears from relaxation of short-range ordered clusters. In principle the short-range order influences the single-particle relaxation process by altering the local single-molecule potential (i.e., the activation

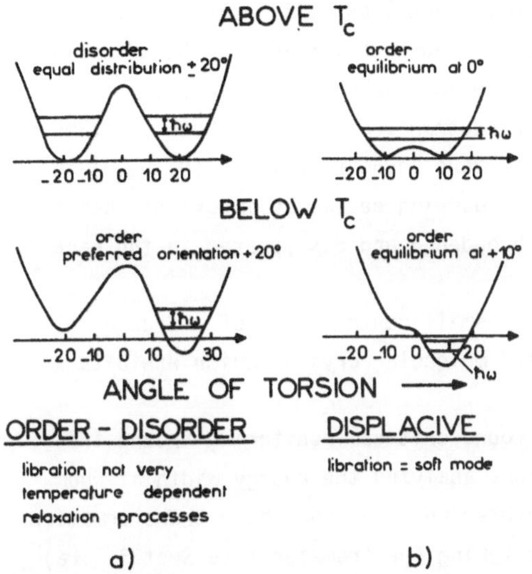

Fig. 3.15a,b. Schematic drawing of the rotational potential and its changes with the phase transformation. (a) For an order-disorder transition; (b) for a displacive transition with a small double-well contribution. One might speculate that (a) corresponds to para-terphenyl and (b) to biphenyl

energy). But this effect is very small. Even the onset of long-range order in CH_4Cl at T_c increases the activation energy only by 13%; see Fig.3.7.

For a displacive phase transformation the bottom of the potential is usually assumed to have a monotonic curvature. But even if there is a small double minimum in the high-temperature phase (see Fig.3.15b) one expects a soft mode as long as the vibration or libration ground state is high compared to the potential barrier. When the soft mode comes down in frequency it gets overdamped sooner or later. Then the lineshape of a heavily overdamped mode cannot be distinguished from a relaxation process of clusters [2.1]. For example, in the paraelectric phase of $BaTiO_3$ it is not clear whether the Ti ions perform large amplitudes around one equilibrium position or whether they are distributed equally over several sites [3.65]. Soft modes and relaxation of clusters are extensively discussed by DORNER and COMES [3.1]. Therefore, only two recently investigated substances from the same chemical family will be presented: biphenyl for the soft mode behavior and para-terphenyl for the disorder type.

Polyphenyls are nonrigid molecules with respect to a torsional angle between the planes of phenyl rings. The orthohydrogen repulsion introduces a nonplanar configuration in the gaseous state. These forces between the phenyl rings are of the same order as the intermolecular forces in the crystalline state which favor the planar configuration. Biphenyl (T_c = 38 K) as well as p-terphenyl (T_c = 178 K) undergo almost continuous phase transformations to antiferrodistortive order of deformed molecules. Recent observations on biphenyl [3.66] revealed a second phase transformation at about 21 K. Both low-temperature structures are incommensurable with the lattice periodicity; for details see Sect.3.4. In the ordered phase the angle between the two phenyl rings of biphenyl is 10° and in p-terphenyl the inner ring is rotated against the outer ones by 21°. In the high-temperature phase the biphenyl appears planar in the structure determination while p-terphenyl exhibits a distribution of the inner ring over two orientations [3.67]. This observation led to the suspicion that p-terphenyl might be of order-disorder type and biphenyl displacive.

CAILLEAU et al. [3.68] performed inelastic neutron scattering on both substances. In biphenyl they observed a soft mode near the zone boundary point B (0, ½, 0) in the temperature range 60 to 200 K. Above 200 K damping became so important that the measured intensity distribution of the mode broadened and disappeared in the background. Below 60 K the mode is overdamped.

It is not excluded that the potential has a small contribution of a double-well potential. Biphenyl is the second investigated molecular crystal which exhibits a soft soft mode after chloranil [3.52].

In p-terphenyl CAILLEAU et al. [3.68,69] found critical scattering around the Brillouin zone boundary point C (½, ½, 0). They analyzed the energy width at room temperature with a normal three axis spectrometer and near the phase transformation temperature with the high-resolution backscattering spectrometer (see Sect.3.3.2a) and found a critical slowing down for the relaxation of the clusters. At about 1.6 K above T_c ($\varepsilon = 0.009$) the lifetime of the clusters was $2.5 \cdot 10^{-9}$ s. This is about one order of magnitude longer as observed in CD_4 [3.70] for a comparable value of ε. After CD_4 p-terphenyl is the second structural order-disorder phase transformation in which critical slowing down could be observed.

In both polyphenyls the phase transformations are caused by the same forces: the orthohydrogen repulsion between phenyl rings. Yet the distortion angle in the ordered phase is only 10 degrees in biphenyl and 21 degrees in p-terphenyl. One might speculate that the larger distortion angle allows a higher potential barrier to build up between the two minima (see Fig.3.15a) thus leading to an order-disorder phase transformation. In biphenyl the smaller distortion angle only permits a small potential barrier (if at all) between the two minima (Fig.3.15b).

Below the phase transformation a soft mode was observed in biphenyl by Raman scattering [3.71], but not in p-terphenyl. For p-terphenyl this is not astonishing as the curvature of the potential minimum should not vary much with temperature.

The more than three-peak structure in the energy distribution of critical soft mode scattering as predicted by AUBRY [3.64] for potentials which have a double-minimum contribution could not be observed up till now.

3.4 Incommensurable Structures

Incommensurable structures (for an introduction see [3.72]) exhibit a pattern of static displacements $\underline{u}$ with a wave vector $\underline{q}_0$ which is not a rational fraction of a reciprocal lattice vector. Under the assumption that the distortion can be described by a single plane wave, we can write

$$\underline{u}(\underline{\ell}) = \underline{A}(T)\cos(\underline{q}_0 \cdot \underline{\ell}) \quad . \tag{3.27}$$

One arbitrary atom (at $\underline{\ell} = 0$) was picked to define an origin. This atom has the static displacement $\underline{u}(0) = \underline{A}(T)$. Incommensurable means that there is no other atom in the originally periodic lattice which has exactly the same static displacement. For a sinusoidal distortion the origin with respect to the lattice is not defined. An arbitrary phase ϕ_0 in the cosine function of (3.27) would not change the energy of the distorted lattice. Thus a continuous change of ϕ_0 would move the displacement pattern through the lattice with no energy needed.

It was this latter aspect, the FRÖHLICH mode [3.73], which stimulated the interest in incommensurable structures especially in conducting materials. A longitudinal distortion of the positions of the positive ions with wave vector $\underline{q}_0$ will be compensated by a charge density wave of the conduction electrons with the same $\underline{q}_0$ as was as well predicted by OVERHAUSER [3.74]. If the distortion can be moved through the crystal with no cost in energy, the electron charge would adiabatically follow. Thus such a system would provide a high-temperature superconductor. But high-temperature superconductors have not yet been observed experimentally, perhaps because defects block the motion of an incommensurate charge density wave.

Detailed theoretical considerations [3.75,76] have shown that the assumption of a sinusoidal wave is too simple. But solutions of the sine-Gordon equation are more realistic. These solutions describe the distortion by solitons (see Sect.3.4.1) which still can move in the ideally pure crystal without cost of energy.

For reviews in this field see [3.2,77-79] and especially for one-dimensional conductors [3.80]. The possibilities of obtaining incommensurable phases are discussed by DVORAK [3.3] in the connected theoretical volume considering the Lifshitz condition.

3.4.1 Amplitude and Phase Modes

A theoretical description of the excitations of the lattice below a phase transformation from a high-symmetry structure to an incommensurable one is difficult, because translational symmetry is lost in the incommensurable phase. The soft mode picture holds quite well for the high-symmetry phase, because in Landau theory the soft mode may appear at any $\underline{q}$.

In the incommensurable phase lattice dynamics in terms of Brillouin zones and normal coordinates of well-defined wave vectors $\underline{q}$ is no longer applicable. But for small incommensurable distortions and, and in order to study the relevant excitations connected with these distortions, one expands the free energy around $\underline{q}_0$. To obtain a translational invariant free energy, one has to introduce coupled modes with wave vectors $\underline{q}_{1,2} = \pm\,\underline{q}_0 + \Delta\underline{q}$. As a solution one finds two contributions to the time dependence of the displacements,

$$\underline{u}_1(\underline{\ell},t) \approx A_1(T)[\exp(i\underline{q}_1 \cdot \underline{\ell}) + \exp(i\underline{q}_2 \cdot \underline{\ell})] \exp(-i\omega_1 t)$$

$$\approx A_1(T) \cos(\underline{q}_0 \cdot \underline{\ell}) \exp[i(\Delta\underline{q} \cdot \underline{\ell} - \omega_1 t)] \quad (3.28)$$

and

$$\underline{u}_2(\underline{\ell},t) \approx \underline{A}_2(T)[\exp(i\underline{q}_1 \cdot \underline{\ell}) - \exp(i\underline{q}_2 \cdot \underline{\ell})] \exp(-i\omega_2 t)$$

$$\approx \underline{A}_2(T) \sin(\underline{q}_0 \cdot \underline{\ell}) \exp[i(\Delta\underline{q} \cdot \underline{\ell} - \omega_2 t)] \quad . \quad (3.29)$$

For $\Delta\underline{q} = 0$ (3.28) describes oscillations in the amplitude of the distortion given by (3.27): "amplitude mode".

For $\Delta\underline{q} = 0$ (3.29) describes phase variations of the distortion or oscillations of the origin: "phase mode" (often called "phason"). The latter can be visualized by

$$\cos(\underline{q}_0 \cdot \underline{\ell} - \phi(t)) = \cos(\underline{q}_0 \cdot \underline{\ell}) + \phi(t) \sin(\underline{q}_0 \cdot \underline{\ell}) \quad . \quad (3.30)$$

The corresponding frequencies are (see Fig.3.16)

$$\omega_1^2 = -2\omega_0^2(T) + \lambda(\Delta q)^2 \qquad \omega_2^2 = \lambda(\Delta q)^2 \quad (3.31)$$

with $\omega_0^2(T) \sim T - T_c$ in Landau theory where T_c is the transition temperature. This description is very simplified because it uses a single plane wave ground state.

The incommensurable wave vector $\underline{q}_0$ usually is close to a commensurable $\underline{q}_c$, where eventually a commensurable superstructure appears after a "lock-in" transition. We can rewrite (3.27)

$$\underline{u}(\underline{\ell}) = \underline{A}(T) \cos[\underline{q}_c \cdot \underline{\ell} + (\underline{q}_0 - \underline{q}_c) \cdot \underline{\ell}] = \underline{A}(T) \cos[\underline{q}_c \cdot \underline{\ell} + \phi(\underline{\ell})] \quad . \quad (3.32)$$

If the distortion were a single plane wave, then ϕ would be a linear function of $\underline{\ell}$ (see dashed line in Fig.3.17) with $|\underline{q}_c| = 1/3 \cdot 2\pi/a$. But as mentioned above, the energy of the distorted crystal can be lowered by introducing solitons.

In this case the distortion is commensurable with the lattice over a distance $r = q_c \cdot a/(q_0 - q_c)$ and then the phase of the distortion jumps rapidly - within a few lattice units - by $q_c \cdot a$ (solid line in Fig.3.17). These jumps in phase are called solitons. The characteristic properties of classical solitons are that they do not dissipate, and that they regain their original shape after scattering from each other. In this sense they are models of elementary particles [3.77]. The number of solitons within a given distance is proportional to $(q_0 - q_c)$.

Neither phase modes nor solitons could as yet be observed near structural phase transformations by inelastic neutron scattering. An exception to the statement about phase modes may be $Hg_{3-\delta}AsF_6$ [3.81]. This compound can be thought of as an

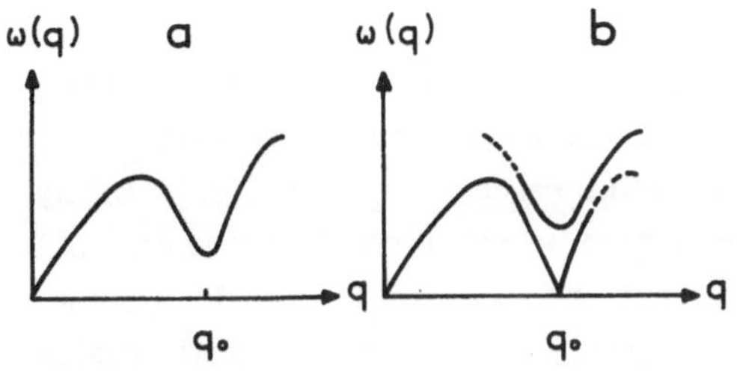

Fig. 3.16a,b

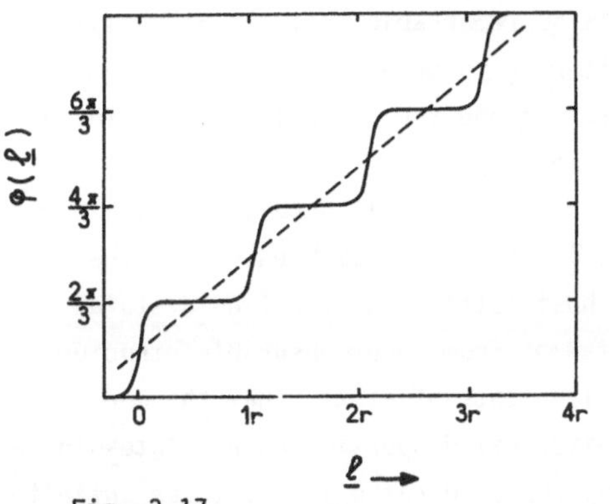

Fig. 3.17

Fig. 3.16a,b. Schematic illustration of the dispersion relation for a material undergoing an incommensurable displacive phase transformation. (a) Above T_0 there is an anomalous dip in the dispersion curve with a soft mode at q_0; (b) below T_0 there appears a splitting into a gapless "phase mode" and an upper "amplitude mode". Pinning of the phase of the static distortion by impurities will cause a gap to appear for the phase mode at q_0 [3.78]

Fig. 3.17. Soliton solutions for the phase of an incommensurable distortion for $q_c = 2\pi/3a$. The dotted line represents the single plane wave approximation [3.75]

ordered tetragonal AsF_6 lattice in which there are open nonintersecting channels running parallel to both basal plane edges. These channels are filled with tightly packed chains of Hg ions. The observed intrachain Hg-Hg distance is such that (3-δ) (δ ≈ 0.18) Hg ions can be accommodated within a unit cell of the AsF_6 lattice. In contrast to the incommensurability of a small displacement wave relative to the underlying lattice, as discussed so far, $Hg_{3-\delta}AsF_6$ features a truly structural incommensurability.

The interaction of the Hg chains with the host lattice is negligibly small. Far above T_c = 120 K the chains can be described as independent one-dimensional liquids [3.82]. Approaching T_c, short-range order builds up within each set of parallel chains. But this short-range order is not a precurser of the low-temperature order. Below T_c, the interaction between the two perpendicular sets of chains produces an incommensurable order between parallel chains of one set such as to make the incommensurable order along one set of chains coincide with the lateral order in the other set and vice versa. Each set of chains produces its characteristic pattern of satellites: $[h(3-\delta), k \pm h\delta, \ell]$ for the chains parallel to the a axis and $[h \pm k\delta, k(3-\delta), \ell]$ for the other set with $h+k+\ell = 2n$. Both patterns have some satellites in common, as, for example, ($h = 1$, $k = 3$, $\ell = 0$) for the first set coincides with ($h = 3$, $k = 1$, $\ell = 0$) for the second set. These experimental observations have been analyzed by a model developed by EMERY and AXE [3.82].

It is remarkable that the Hg subsystem above T_c does not exhibit truly elastic scattering to be described by a δ function in energy. This means that the Hg ions diffuse along the channels like a liquid and the infinite time correlations tend to zero.

The longitudinal sound velocity along the chains was measured above T_c (liquid phase). This mode has been called the phase mode, but the negligible coupling to the host lattice and the liquid status of the chains render the system sufficiently different from incommensurable structures consisting of small static incommensurable displacements of atoms, that the word phase mode is not applicable. For $T > T_c$, the longitudinal dispersion extrapolates to zero frequency (within experimental resolution) for $q \to 0$ (at q_0), i.e., for infinite wavelength of the longitudinal acoustic phonon (phase mode) in the Hg chains. This means just easy Hg mass transport along the channels in the host crystal, an effect which has been observed in the crystal-growing technique.

As far as solitons are concerned, they have been observed by inelastic neutron scattering so far only in the one-dimensional ferromagnet $CsNiF_3$ [3.83] as predicted by MIKESKA [3.84].

In most substances the incommensurable wave vector $\underline{q}_0$ is temperature dependent. With decreasing temperature $\underline{q}_0$ approaches $\underline{q}_c$ by changing its length, for example, in 2H - $TaSe_2$ [3.85] or its direction as, for example, in $1T_2$ - TaS_2 [3.86].

Many substances show a "lock-in" transition from the incommensurable to the commensurable structure. This lock-in transition was in all cases studied experimentally found to be a first-order one. MONCTON et al. [3.85] developed a free energy in Landau theory for 2H - $TaSe_2$. Their theory predicts a step in $q_0 - q_c$ even bigger than observed. This theory goes already beyond the concept of a single plane wave ground state as it includes two waves: one at $q_0 = q_c - \delta$ and one at $q' = q_c + 2\delta$ where $\delta = q_c - q_0$. In the incommensurable structure two different satellites were observed at $q_c - \delta$ and $q_c + 2\delta$ (see Fig.3.18).

In acceptable agreement with simplest theory, the intensity of the $q_c + 2\delta$ satellite was proportional to the square of the intensity of the $q_c - \delta$ satellite. The amplitude of the $q_c - \delta$ distortion is the primary order parameter.

3.4.2 Conducting Materials

The electron-phonon interaction generally leads to Kohn anomalies [3.87] which are more or less visible as anomalies in the phonon dispersion curves. The renormalization of the phonon frequency Ω due to the interaction with the electrons is given by

$$\omega^2(\underline{q}) = \Omega^2(\underline{q})[1 - \lambda(\underline{q})\chi(\underline{q})] \quad , \tag{3.33}$$

where $\lambda(\underline{q})$ is the electron-phonon coupling constant and $\chi(\underline{q})$ the static electron-gas susceptibility

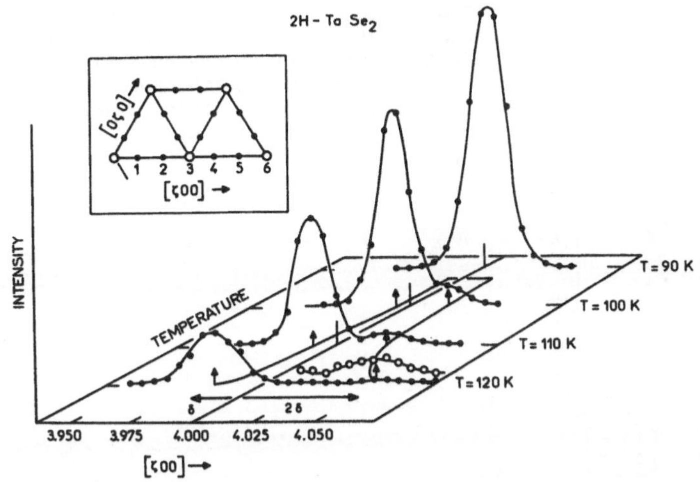

Fig. 3.18. Intensity profiles of the incommensurate Bragg satellites in 2H - $TaSe_2$. In addition to the primary satellite at $\zeta = 4 - \delta$, there appears a secondary peak at $\zeta = 4 + 2\delta$, resulting from the nonsinusoidal nature of the distortion. The inset shows indexing convention in the basal plane in the commensurate phase. The open circles represent parent reciprocal lattice positions, the closed circles the superlattice positions [3.85]

$$\chi(\underline{q}) = \sum_{\underline{k}} \frac{f(E_{\underline{k}})-f(E_{\underline{k}+\underline{q}})}{E_{\underline{k}+\underline{q}}-E_{\underline{k}}} \quad . \tag{3.34}$$

Here $E_{\underline{k}}$ is the energy of the electronic state of momentum $\underline{k}$ and $f(E_{\underline{k}})$ is the Fermi distribution function.

Anomalies in $\chi(\underline{q})$ will arise from electron-electron scattering processes with maximal $\underline{q}$. These are scattering processes between opposite sides of the Fermi surface with a wave vector $\underline{q}$ equal to the distance between the two sides, most often $\underline{q} = 2\,\underline{k}_F$. The strength of the anomaly depends on the shape of the Fermi surface, in other words on the number of electrons having almost the same $\underline{k}$ around $\underline{k}_F$. In the three-dimensional case (normal metals) the Fermi surface for free electrons is a sphere: the anomalies are weak; in the two-dimensional case (e.g., transition-metal dichalcogenides) the Fermi surface for free electrons is a cylinder: anomalies in the phonon dispersion and incommensurate phase transformations at $q_0 = 2\,\underline{k}_F$ have been observed (Peierls transitions [3.88]); in the one-dimensional case (e.g., KCP, TTF-TCNQ) the Fermi surface is a pair of planes: strong anomalies in the phonon dispersion and incommensurate phase transformations at $\underline{q}_0 = 2\,k_F$ have been observed [3.80], even satellites at $\underline{q}_0' = 4\,k_F$ appeared.

The anomalies (dips) in the phonon dispersion curves above the phase transformation depend only weakly on temperature. Figure 3.19 shows the anomaly in the

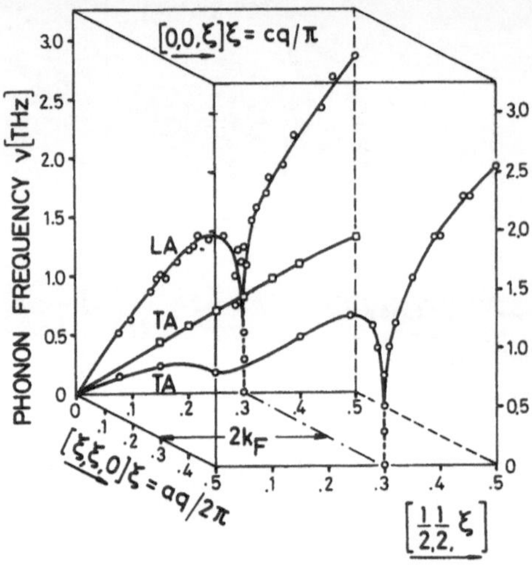

Fig. 3.19. Phonon dispersion in $K_2Pt(CN)_4Br_{0.3} \times H_2O$ at 300 K. The dip in the dispersion corresponds to a giant Kohn anomaly which appears in planes perpendicular to the z axis and at a distance of 2 k_F from the origin [3.89]

phonon dispersion of $K_2Pt(CN)_4\ Br_{0.30} - xH_2O$ (KCP) as measured by RENKER et al. [3.89]. The narrow deep valley appears everywhere in a plane in reciprocal space. This plane is perpendicular to the conducting Pt chains and corresponds to a planar Fermi surface.

A soft mode behavior with a phonon mode approaching zero frequency as $T \to T_c$ was never observed in a conductor even if the incommensurable transition is almost second order. Lock-in transitions where $|q_0| = |2\ \underline{k}_F| \to q_c$ have only been observed in two-dimensional systems like 2H - $TaSe_2$ [3.85] because here the Fermi surface can be deformed. A cylinder with circular cross section can get an elliptic cross section. In one-dimensional systems $|\underline{q}_0| = |2\underline{k}_F|$ was found temperature independent because there is no way to vary $|\underline{k}_F|$. But other incommensurable components q_0' appear perpendicular to the conducting chains from interchain ordering, which exhibit lock-in transitions [3.76].

A very interesting parameter to vary is pressure. Atomic distances vary with it and thus the interactions. Even the conduction electron density may depend on pressure so that $|\underline{k}_F|$ may change. Studies on TTF - TCNQ under hydrostatic pressure have been started recently [3.90]. As there exists a recent review on one-dimensional conductors by COMES and SHIRANE [3.80], we have kept this section short.

3.4.3 Insulating Materials

It might seem surprising that incommensurable structures appear as well in insulators, where an unstable electronic system is absent. Therefore it must be the lattice itself which becomes unstable. But the only necessary condition for a soft mode phase transformation to occur is an extremum (usually a minimum) in the phonon dispersion surface: a van Hove critical point. Certain van Hove singularities occur necessarily at high-symmetry points in the Brillouin zone, and this accounts in a general way for the frequent occurrence of Brillouin zone center and zone boundary soft mode instabilities. Whether and where other minima appear depends entirely on the strength and the range of the interatomic potential.

Note that (in nonsymmorphic space groups) for symmetry directions which contain operations such as screw axes or glide planes, the zone boundary is not a van Hove critical point, if the periodicity distance in real space for this symmetry direction is enlarged by the presence of the symmetry operation containing a translation. Then two modes which are degenerate at the zone boundary come in with opposite but finite slope. Therefore a presentation of the dispersion curves in the extended zone scheme (see Fig.3.20) for K_2SeO_4 [3.91] is not only convenient, but displays the physical connection between acoustic (left-hand side) and optic modes (right-hand side).

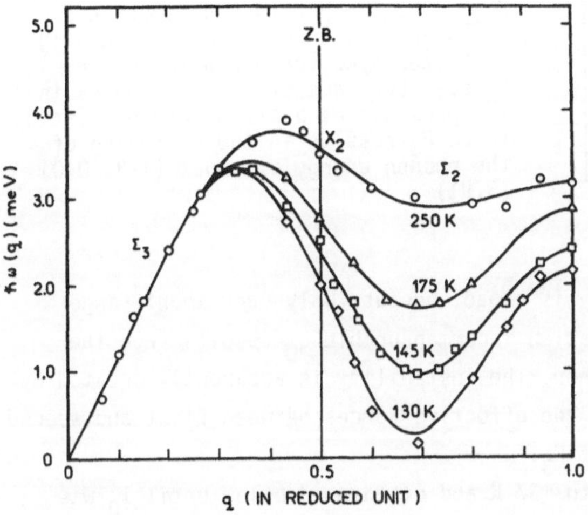

Fig. 3.20. Phonon dispersion of the Σ_3 (acoustic) and the Σ_2 (soft optic mode) of K_2SeO_4 at different temperatures in an extended zone scheme [3.91]

K_2SeO_4 has three phase transformations: from hexagonal to orthorhombic at 745 K; from orthohombic to incommensurable at T_0 = 130 K (second order); and from incommensurable to ferroelectric commensurable at T_c = 93 K (first order). The orthohombic phase is designated P (paraelectric), the incommensurable I with satellites at $q_0 = (1-\delta)2\pi/3a$ ($\delta \approx 0.07$ at 122.5 K), and the lowest F (ferroelectric in c direction) associated with a tripling of the unit cell in a direction, which is parallel to the high-temperature hexagonal axis.

A broad anomaly in an optic phonon dispersion branch with a soft mode at the minimum was observed in phase P (see Fig.3.20). The anomaly in KCP (see Fig.3.19) was much narrower because the anomaly in the conduction electron susceptibility is confined to the immediate neighborhood of $2k_F$.

IIZUMI et al. [3.91] analyzed the dispersion of the anomaly by effective forces F_n between first, second, ... sixth neighboring layers, which are separated by $n \cdot a/2$, in terms of the Fourier series

$$\omega^2(q) = \sum_n F_n[1 - \cos(nq)] \quad . \tag{3.35}$$

The least squares fitted curves for different temperatures are shown in Fig.3.20. The coefficients F_n for $n > 3$ turned out to be insignificant. The temperature dependence of the first three coefficients as obtained from the fits is given in Fig.3.21.

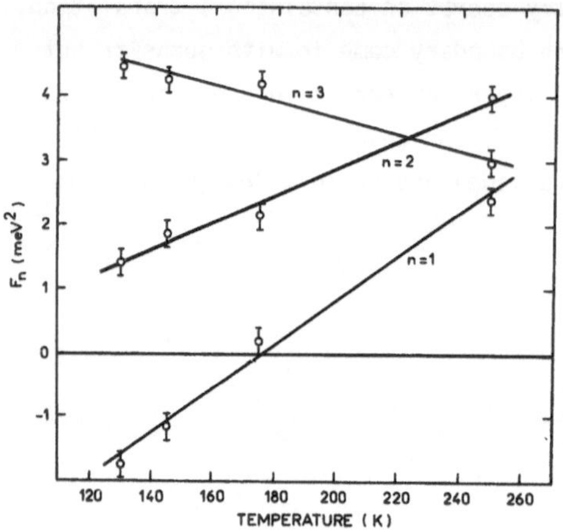

Fig. 3.21. Temperature dependence of the nearest-, second-, and third-neighbor interlayer force constants for K_2SeO_4. Decrease of both F_1 and F_2 in the presence of a strong and persisting force F_3 results in the softening of the phonon energy at about (1/3, 0,0) [3.91]

Now one understands that the anomaly is broad, because only very short-range interactions play a role. The minimum at an incommensurable q_0 results from the values of the F_n' s relative to each other. The instability is apparently created by competing forces, here the decrease of the effective forces between first and second neighboring planes and the constancy of F_3.

The soft mode was underdamped down to 137 K and overdamped below until T_0 was reached. A central peak was observed below 137 K. Its intensity diverges at T_0. The integrated intensity I_{int} (over central peak and soft mode) had the temperature dependence expected from molecular field theory as measured by X-ray scattering [3.92]

$$I_{int}/T \sim \omega_0^{-2} \sim (T - T_0)^{-1} \quad . \tag{3.36}$$

IIZUMI et al. [3.91] produced a free energy within Landau theory which could describe the important features of the phases I and F. The primary order parameter η_δ is the amplitude of the incommensurable distortion at $q_0 = (1 - \delta)2\pi/3a$. Third order was the lowest order in which a secondary order parameter $\eta_{3\delta}$ could couple. η_{3d} is the amplitude of electric polarization in c direction and is proportional to η_δ^3. To maintain translational invariance in the free energy, the wave vector $q_{3\delta}$ corresponding to the secondary order parameter has to be

$$q_{3\delta} = 2\pi/a - 3q_0 = \delta \cdot 2\pi/a \quad . \tag{3.37}$$

This is a small wave vector near the Brillouin zone center. Despite the experimental difficulties intensity from this satellite could be measured.

There is no macroscopic spontaneous polarization in phase I, because the amplitude of the polarization $\eta_{3\delta}$ is modulated by the wave $q_{3\delta}$. At the lock-in transition at T_c, δ drops to zero and with it $q_{3\delta} = 0$. Then the amplitude $\eta_{3\delta}$ which has built up in phase I manifests itself as a finite macroscopic polarization.

Minimizing the free energy with respect to η_δ, $\eta_{3\delta}$, and δ the temperature dependence of δ could be described including the step at T_c due to the locking in of the commensurable distortion, in good agreement with the experimental results.

Very recently two incommensurate phases have been observed in biphenyl [3.66]. Biphenyl is monoclinic with the nonsymmorphic space group $P2_{1/a}$ in phase I. At $T_I = 38$ K the incommensurate phase II appears with satellites at

$$q_S = \delta_a \cdot \underline{a}^* + \tfrac{1}{2}(1 - \delta_b)\underline{b}^* \quad , \qquad (3.38)$$

where $\underline{a}^*$ and b* are reciprocal lattice vectors.

Altogether there appear four satellites around the Brillouin zone boundary point B = (0, ½, 0) (see Fig.3.22). $\delta_a \approx 0.05$ and $\delta_b \approx 0.075$ decrease little with temperature. In phase III, below $T_{II} = 21$ K only the incommensurability along b is left (see Fig.3.22). At T_{II} δ_a drops to zero and δ_b decreases abruptly by about 0.005. The temperatures T_I and T_{II} correspond to the deuterated compound. For the non-deuterated one they are $T_I = 40$ K and $T_{II} = 16$ K.

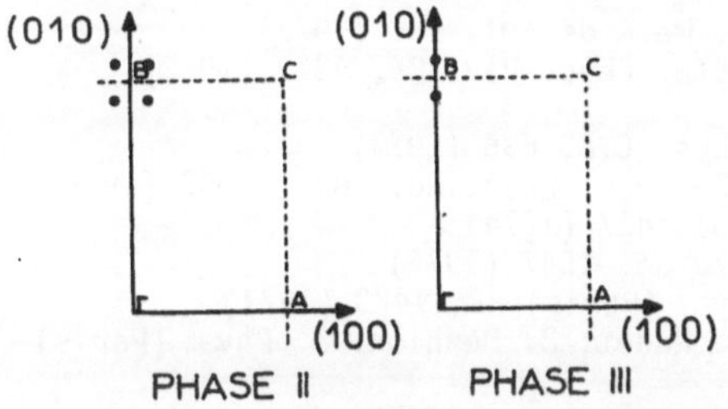

Fig. 3.22. Locations of satellite reflections of biphenyl in the (hk0) scattering plane for phase II and phase III. Dotted lines correspond to the Brillouin zone boundaries [3.66]

The investigation of excitations around the satellites in the incommensurate phase with inelastic neutron scattering is underway. Rb_2ZnBr_4 [3.93] has an incommensurable phase for $200\text{ K} < T < 355$ K. In this substance the soft mode is overdamped above 355 K; therefore a frequency of the soft mode could not be determined. The amplitude and phase fluctuations in the incommensurate phase were overdamped as well. They are hidden in quasielastic diffuse scattering.

Much progress in the study of excitations in incommensurable strucutres is expected for the near future.

As outlined in this chapter, inelastic scattering of neutrons is a very suitable technique to study correlations in space and time. But the limits are quite often set by the available flux. More flux would be desired to study small samples and to improve resolution. For continuous neutron sources any progress beyond the existing high flux beam reactors will be costly and slow.

I would like to thank R. Currat, A. Heidemann, A. Hüller, R. Lechner, R. Pynn, and W.G. Stirling for helpful discussions and carefully reading the manuscript.

References

3.1 B. Dorner, R. Comes: In *Dynamics of Solids and Liquids by Neutron Scattering*, Topics in Current Physics, Vol. 3, ed. by S.W. Lovesey, T. Springer (Springer, Berlin, Heidelberg, New York 1977)
3.2 R. Currat, R. Pynn: *Neutron Scattering in Materials Science*, ed. by G. Kostorz in the series "A Treatise on Materials Science and Technology" (Academic Press, New York 1979)
3.3 V. Dvorak: *Symmetry Aspects of Structural Phase Transformations* in Topics in Current Physics, ed. by H. Thomas, K.A. Müller
3.4 H. Thomas, H. Hoeck: *Structural Phase Transformations* in Topics in Current Physics, ed. by H. Thomas, K.A. Müller
3.5 W. Cochran: Phys. Rev. Lett. *3*, 412 (1959)
3.6 P.W. Anderson: In *Fizika dielektrikov*, ed. by G.I. Skanavi (Acad. Nauk SSSR, Moscow 1960, p.290)
3.7 J.D. Axe, G. Shirane, K.A. Müller: Phys. Rev. *183*, 820 (1969)
3.8 T. Riste, E.J. Samuelsen, K. Otnes, J. Feder: Solid State Commun. *9*, 1455 (1971)
3.9 J.K. Kjems, G. Shirane, K.A. Müller, H.J. Scheel: Phys. Rev. B*8*, 1119 (1973)
3.10 K.A. Müller: In *Dynamical Critical Phenomena and Related Topics*, Lecture Notes in Physics, Vol. 104, ed. by C.P. Enz (Springer, Berlin, Heidelberg, New York 1979) p.210
3.11 S.M. Shapiro, N.J. Chesser: Nucl. Instrum. Methods *101*, 183 (1972)
3.12 S.M. Shapiro, J.D. Axe, G. Shirane, T. Riste: Phys. Rev. B*6*, 4332 (1972)
3.13 R.A. Cowley: Ferroelectrics *6*, 163 (1974)
3.14 J. Töpler, B. Alefeld, A. Heidemann: J. Phys. C*10*, 635 (1977)
3.15 R. Currat, K.A. Müller, W. Berlinger, F. Denoyer: Phys. Rev. B*17*, 2937 (1978)
3.16 A. Aharony, A.D. Bruce: Phys. Rev. Lett. *33*, 427 (1974)
3.17 K.A. Müller, W. Berlinger: Phys. Rev. Lett. *35*, 1547 (1975)
3.18 M. Rousseau, J. Nouet, R. Almairac: J. Phys. (Paris) *38*, 1423 (1977)
3.19 R. Almairac, M. Rousseau, J.Y. Gesland, J. Nouet, B. Hennion: J. Phys. (Paris) *38*, 1429 (1977)
3.20 F. Denoyer, R. Currat: In *Neutron Inelastic Scattering* (IAEA, Vienna 1978) p.278
3.21 B.I. Halperin, C.M. Varma: Phys. Rev. B*14*, 4030 (1976)
3.22 J.D. Axe, S.M. Shapiro, G. Shirane, T. Riste: In *Anharmonic Lattices, Structural Transitions and Melting*, ed. by T. Riste (Noordhoff, Leiden 1974) p.23
3.23 J.B. Hastings, S.M. Shapiro, B.C. Frazer: Phys. Rev. Lett. *40*, 237 (1978)
3.24 D. Bäuerle, W. Rehwald: Solid State Commun. *27*, 1343 (1978)
3.25 D. Wagner, D. Bäuerle, F. Schwabl, B. Dorner, H. Kraxenberger: Z. Phys. B*37*, 317 (1980)
D. Wagner, D. Bäuerle, F. Schwabl, M. Wöhlecke, B. Dorner, H. Kraxenberger: Ferroelectrics *26*, 725 (1980)
3.26 J.M. Rowe, J.J. Rush, N.H. Chesser, K.H. Michel, J. Neudts: Phys. Rev. Lett. *40*, 455 (1978)
3.27 K.A. Müller, W. Berlinger, C.H. West, P. Heller: Phys. Rev. Lett. *32*, 160 (1974)
3.28 G.F. Reiter, W. Berlinger, K.A. Müller, P. Heller: To be published (1980)

3.29 P.A. Fleury, K.B. Lyons: Chap. 2 in this book
3.30 A.D. Bruce, K.A. Müller, W. Berlinger: Phys. Rev. Lett. *42*, 185 (1979)
3.31 W. Press, A. Hüller: Acta Crystallogr. A*29*, 257 (1973)
3.32 A.J. Leadbetter, R.E. Lechner: In *The Plastically Crystaline State*, ed. by J.N. Sherwood (Wiley, New York 1979) p.285
3.33 J.D. Barnes: In *Neutron Inelastic Scattering 1972* (IAEA, Vienna 1972) p.287
3.34 J. Töpler, D.R. Richter, T. Springer: J. Chem. Phys. *69*, 3170 (1978)
3.35 Scientific Secretary ILL: "*Neutron Beam Facilities at the HFR Available for Users*" (Institut Laue-Langevin, 156X, 38042 Grenoble Cedex, France, 1977)
3.36 B. Jacrot: In *Instrumentation for Neutron Inelastic Scattering Research* (IAEA, Vienna 1970) p.225
3.37 M. Birr, A. Heidemann, B. Alefeld: Nucl. Instrum. Methods *95*, 435 (1971)
3.38 B. Alefeld: Kerntechnik *14*, 15 (1972)
3.39 F. Douchin, R.E. Lechner, R. Scherm: Nucl. Instrum. Methods, to be published (1980)
3.40 F. Mezei: In *Neutron Inelastic Scattering 1977* (IAEA, Vienna 1978) p.125
3.41 W. Press: J. Chem. Phys. *56*, 2597 (1972)
3.42 H. Grimm, W. Press: Private communication (1977)
3.43 H. Kapulla, W. Gläser: In *Neutron Inelastic Scattering 1972* (IAEA, Vienna 1972) p.841
3.44 W. Press, A. Kollmar: Solid State Commun. *17*, 405 (1975)
3.45 W. Rehwald, J.R. Sandercock, M. Rossinelli: Phys. Status Solidi (a) *42*, 699 (1977)
3.46 H.C. Teh, B.N. Brockhouse: Phys. Rev. B*3*, 2733 (1971)
3.47 E.R. Cowley: Phys. Rev. B*3*, 2743 (1971)
3.48 G. Venkataraman, K. Usha, P.R. Iyengar, P.R. Vijayaraghavan, A.P. Roy: In *Inelastic Scattering in Solids and Liquids*, Vol.II (IAEA, Vienna 1963) p.253
3.49 J.C. Damien, R. Currat, J. Lefebvre, B. Hennion: In *Neutron Inelastic Scattering 1977* (IAEA, Vienna 1978) p.331
3.50 J.C. Damien, J. Lefebvre, M. More, B. Hennion, R. Currat: J. Phys. C*11*, 4323 (1978)
3.51 W.G. Stirling, W. Press, H.H. Stiller: J. Phys. C*10*, 3959 (1977)
3.52 W.D. Ellenson, J.K. Kjems: J. Chem. Phys. *67*, 3619 (1977)
3.53 K.H. Michel: J. Chem. Phys. *58*, 1143 (1973)
3.54 R.C. Livingston, J.M. Rowe, J.J. Rush: J. Chem. Phys. *60*, 4541 (1974)
3.55 A. Hüller, J.W. Kane: J. Chem. Phys. *61*, 3599 (1974)
3.56 R. Stockmeyer, H.H. Stiller: Phys. Status Solidi *27*, 169 (1968)
3.57 R.E. Lechner, A. Heidemann: Commun. Phys. *1*, 213 (1976)
3.58 R.E. Lechner: In *Neutron Scattering*, ed. by R.M. Moon, Conference Gattlinburg, CONF-760601-81, Distribution Category NC-34 (1976)
3.59 Y. Yamada, Y. Noda, J.D. Axe, G. Shirane: Phys. Rev. B*9*, 4429 (1974)
3.60 S. Haussühl: Solid State Commun. *13*, 147 (1973)
3.61 J. Daubert, K. Knorr, W. Dultz, H. Jex, R. Currat: J. Phys. C*9* L 389 (1976)
3.62 K.H. Michel, J. Naudts, B. De Raedt: Phys. Rev. B*18*, 648 (1978)
3.63 K.H. Michel, J. Naudts: J. Chem. Phys. *67*, 547 (1977), *68*, 216 (1978)
3.64 S. Aubry: J. Chem. Phys. *62*, 3217 (1975)
3.65 R. Comès, M. Lambert, A. Guinier: Acta Crystallogr. A*26*, 244 (1970)
3.66 H. Cailleau, F. Moussa, J. Mons: Solid State Commun. *31*, 521 (1979)
3.67 J.L. Baudour, H. Cailleau, W.B. Yelon: Acta Crystallogr. B*33*, 1773 (1977)
3.68 H. Cailleau, A. Girard, F. Moussa, C.M.E. Zeyen: Solid State Commun. *29*, 259 (1979)
3.69 H. Cailleau, A. Heidemann, C.M.E. Zeyen: J. Phys. C*12*, L 411 (1979)
3.70 W. Press, A. Hüller, H.H. Stiller, W.G. Stirling, R. Currat: Phys. Rev. Lett. *32*, 1354 (1974)
3.71 A. Bree, M. Edelson: Chem. Phys. Lett. *46*, 500 (1977)
3.72 R. Pynn: Nature *281*, 433 (1979)
3.73 H. Fröhlich: Proc. R. Soc. London A*223*, 296 (1954)
3.74 A.W. Overhauser: Phys. Rev. *167*, 691 (1968)
3.75 W.L. McMillan: Phys. Rev. B*14*, 1496 (1976)
3.76 P. Bak, V.J. Emery: Phys. Rev. Lett. *36*, 978 (1976)

3.77 V.J. Emery: In *Chemistry and Physics of One Dimensional Metals*, ed. by H.J. Keller (Plenum, New York, London 1977) p.1
3.78 J.D. Axe: In *Neutron Scattering*, - Proceedings of the Conference in Gatlinburg, ed. by R.M. Moon (National Technical Information Service, U.S. Department of Commerce, Springfield, Virginia 22161, 1976) p.353
3.79 J.D. Axe: In *Neutron Inelastic Scattering 1977* (IAEA, Vienna 1978) p.101
3.80 R. Comès, G. Shirane: In *Highly Conducting One Dimensional Solids*, ed. by G.T. Devreese (Plenum, New York 1979) p.17
3.81 I.U. Heilmann, J.D. Axe, J.M. Hastings, G. Shirane, A.J. Heeger, A.G. MacDiarmid: Phys. Rev. B*20*, 751 (1979)
3.82 V.J. Emery, J.D. Axe: Phys. Rev. Lett. *40*, 1507 (1978)
3.83 M. Steiner, J.K. Kjems: Phys. Rev. Lett. *41*, 1137 (1978)
3.84 H.J. Mikeska: J. Phys. C*11*, L 29 (1978)
3.85 D.E. Moncton, J.D. Axe, F. DiSalvo: Phys. Rev. B*16*, 801 (1977)
3.86 C.B. Scruby, P.M. Williams, G.S. Parry: Philos. Mag. *197*, 255 (1975)
3.87 W. Kohn: Phys. Rev. Lett. *2*, 393 (1959)
3.88 R.E. Peierls: *Quantum Theory of Solids* (Clarendon, Oxford 1964)
3.89 B. Renker, L. Pintschovius, W. Gläser, H. Rietschel, R. Comès: In *Laser Spectroscopy*, Lecture Notes in Physics, Vol.43 (Springer, Berlin, Heidelberg, New York 1975) p.53
3.90 S. Megtert, R. Pynn, C. Vettier, R. Comès, A.F. Garito: "Order in Strongly Fluctuating Condensed Matters Systems" - Proceedings of NATO ASI, Geilo (1979)
3.91 M. Iizumi, J.D. Axe, G. Shirane: Phys. Rev. B*15*, 4392 (1977)
3.92 H. Terauchi, H. Takenaka, K. Shimaoka: J. Phys. Soc. Jpn. *39*, 436 (1975)
3.93 C.J. de Pater, J.D. Axe, R. Currat: Phys. Rev. B*19*, 4684 (1979)

Additional References with Titles

Aubry S: The New Concept of Transitions by Breaking of Analyticity in a Cristallographic Model, in *Solitons and Condensed Matter Physics*, ed. by A.R. Bishop, T. Schneider, Springer Series in Solid-State Sciences, Vol.8 (Springer, Berlin, Heidelberg, New York 1978) p.264-277

Cailleau H., Moussa F., Zeyen C.M.E., Bouillot J: Observation of excitations in the incommensurable phases of biphenyl by inelastic neutron scattering. Solid State Commun. *33*, 407-411 (1980)

Denoyer F., Moudden A.H., Lambert M: Neutron scattering study of the incommensurate phase transition in thiourea. Ferroelectrics *24*, 43-51 (1980)

Joffrin C., Benoit J.P., Currat R., Lambert M: Inelastic Neutron scattering study of the ferroelectric phase transition in $Pb_3(PO_4)_2$. J. Phys. (France) *40*, 1185-1194 (1979)

Lefebvre J., Currat R., Fouret R., More M: Neutron diffusion studies of lattice vibrations in sodium nitrate. J. Phys. C*13*, 4449-4461 (1980)

Lowde R.D., Harley R.T., Saunders G.A., Sato M., Scherm R., Underhill C: On the martensitic transformation in FCC manganese alloys, I - measurements. Proc. Roy. Soc. (in press, 1980)

Moudden A.M., Denoyer F., Lambert M., Fitzgerald W: On the lock - in phase transformation in thiourea. Solid State Commun. *33*, 933-936 (1979)

Press W., Majkrzak C.F., Axe J.D., Hardy J.R., Massa N.E., Ullmann F.G: Effect of hydrostatic pressure on the incommensurate phase of K_2SeO_4. Phys. Rev. B*22*, 332-335 (1980)

Press W., Prager M., Heidemann A: Rotational tunneling in methane (CD_4): isotope effect. J. Chem. Phys. *72*, 5924-5926 (1980)

Press W., Renker B., Schulz H., Böhm H: Neutron scattering study of the one-dimensional ionic conductor β-eucryptite. Phys. Rev. B*21*, 1250-1257 (1980)

Sato M., Lowde R.D., Saunders G.A., Hargreave M.M: On the martensitic transformation in FCC manganese alloys, II - phenomenological analysis. Proc. Roy. Soc. (in press, 1980)

4. Ultrasonic Studies Near Structural Phase Transitions

B. Lüthi and W. Rehwald

With 11 Figures

In this chapter we discuss elastic properties of solids near structural phase transitions. It will be shown that by the measurement of the elastic response important information in the field of phase transitions can be obtained. For example, elastic constant measurements are a very sensitive tool to locate transition points, to determine phase diagrams, and, in favorable cases, to make statements about the order of the phase transition. Furthermore, from the temperature variation of the elastic response functions the type of coupling between strain and order parameter can be deduced. Based on its knowledge, the coupling parameters and the characteristic variations of other important quantities can be derived. In addition, elastic measurements are used to test aspects of the modern renormalization theories, such as crossover and dimensional effects.

Most experiments described in this chapter can be performed with relatively simple and inexpensive equipment, as will be discussed below, but the information gained is quite substantial. Ultrasonics is therefore an attractive method to be used in the field of phase transitions. This we would like to demonstrate in this chapter.

4.1 Background Material

We start with the definitions of the elastic response functions and a short listing of experimental methods to determine them.

4.1.1 Elastic Response Functions and Their Measurement

In the frame of linear elasticity theory the components of the stress tensor $\underline{\underline{\sigma}}$ and of the strain tensor $\underline{\underline{e}}$ are linearely related,

$$\sigma_{ij} = \sum_{k\ell} c_{ijk\ell} e_{k\ell}$$

$$e_{ij} = \sum_{k\ell} s_{ijk\ell} \sigma_{k\ell} \quad . \qquad (4.1)$$

These equations define elastic stiffness functions $c_{ijk\ell}$, sometimes called elastic constants or elastic stiffnesses, and compliances $s_{ijk\ell}$, either of which can be used to characterize the elastic response. Generally a contracted Voigt notation is used,

$$ij \to m = i \quad \text{if} \quad i = j \quad \text{and} \quad ij \to m = 9 - i - j \quad \text{if} \quad i \neq j \quad .$$

Only in a static load experiment are the functions c and s determined directly. In most cases they are obtained via measurements of the sound velocities v_s. When dissipative processes are also considered, the elastic response functions become complex, and their imaginary part characterizes mainly the ultrasonic attenuation α_s.

a) *Pulse Method*

A simple way to measure both quantities, v_s and α_s, in one experiment is offered for example by the pulse echo method as shown schematically in Fig.4.1. By a piezoelectric transducer, bonded to the crystal investigated, a short pulse of sound waves is generated. If the sample has a pair of end faces, plane and parallel, the train of sound waves is multiply reflected and produces an electric signal every time it hits the transducer. These electric echo pulses are amplified and displayed on an oscilloscope or processed otherwise. From the length, the transit time, and the decay rate the velocity and attenuation can be obtained immediately. Such measurements give the absolute velocity with an accuracy of about 1%.

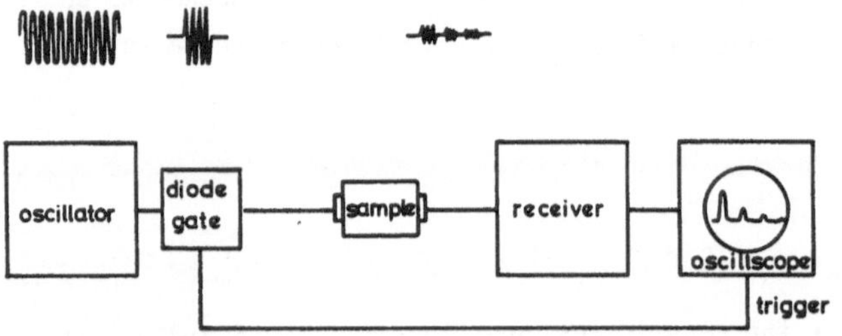

Fig.4.1. Pulse echo apparatus

Velocity changes can be measured with much higher precision, using *phase sensitive methods*. Various methods are employed: pulse superposition [4.1], phase comparison [4.2], sing-around method [4.3], and pulse overlap method [4.4]. They all permit measurement of small variations in the sound velocity of the order of 10^{-4} even in attenuating materials and down to 10^{-7} in favorable cases. For more details on the experimental method we refer to specialized review articles [4.5-7].

b) *CW Methods*

Standing wave or cw methods have also been successfully applied to various problems in physical acoustics. Similar to a Fabry-Perot interferometer one excites standing wave resonances generally also with quartz transducers. For a sample of length L the number of excited resonances of frequency f is $n = 2Lf/v$ whereby the sound velocity v can be determined. For 10 MHz n is of the order of 10^2. Using frequency modulation techniques one can measure changes in velocity and attenuation with high precision. For a detailed discussion we refer to review articles [4.8,9].

In such ultrasonic experiments, both pulse and cw methods, the frequency is usually in the range of 10 to 100 MHz. This means that the sound wavelength is short compared to the lateral sample dimensions and v_s and α are connected with the corresponding stiffness c_s,

$$q^* = \frac{\omega_s}{v_s} - i\alpha_s = \omega\left(\frac{\rho}{\mathrm{Re}\{c_s\} + i\mathrm{Im}\{c_s\}}\right)^{\frac{1}{2}} \approx \omega\left(\frac{\rho}{\mathrm{Re}\{c_s\}}\right)^{\frac{1}{2}} - i\,\frac{\omega \mathrm{Im}\{c_s\}}{2\rho v_s^3} \quad . \tag{4.2}$$

The approximation is valid for $\mathrm{Im}\{c_s\} \ll \mathrm{Re}\{c_s\}$, a condition usually fulfilled. The mass density is denoted by ρ.

c) *Low-Frequency Methods*

The lower limit of frequency is given by the sample dimensions. Here the number n of mechanical resonances is of order unity. In this case the elastic compliances (Young's modulus Y and shear modulus G) are determined by a cw resonance method or by measuring flexural and torsional oscillations. In the cubic system $Y = 1/s_{11}$ and $G = 1/s_{44} = c_{44}$ hold. These techniques were described by READ et al. [4.10]. These methods are particularly suitable for piezoelectric materials, which can be excited into mechanical resonances by an electric field directly without transducers [4.11]. Other nonpiezoelectric crystals can be measured in a similar way with an additional dc bias field using the electrostrictive effect [4.12].

The upper limit in frequency in an ultrasonic experiment is given by the precision to which planeness and parallelism of the two reflecting end faces can be achieved. For coherent detection this precision has to be about 1/10 of the acoustic wavelength. For high acoustic quality materials one can go well into microwave frequencies. A complementary tool in this frequency region is Brillouin scattering, which is described in a separate chapter. Some results of this technique will be used to supplement data from ultrasonic investigations.

A serious problem for ultrasonic propagation near structural phase transition is the bonding of transducers to the sample. Because of thermal expansion and the occurrence of spontaneous strains in the low symmetry phase the transducer sample bond may crack under such conditions. With quartz and ceramic PZT transducers one generally uses as bonding materials Dow-Corning 806 A resin for high temperatures and Dow-Corning RTV silicone or nonaq stopcock grease for low temperatures. A novel technique is the direct evaporation of CdS or ZnO transducers onto the sample, avoiding these bonding problems altogether [4.13]. This technique has been used predominantly for high-frequency acoustics (500 MHz and higher), but it can be also applied down to 10 MHz [4.14].

There are several other special techniques being used for ultrasonic propagation studies near structural phase transitions. We would like to mention two of them.

By the so-called "phonon echoes" the ultrasonic attenuation can be measured in irregularly shaped samples (e.g., [4.15]). Combining a phonon echo active material (e.g., $Bi_{12}GeO_{20}$) with the crystal under investigation, accurate attenuation studies are possible [4.15].

An old method, which works only for transparent materials, is the diffraction of light by ultrasonic waves [4.16,17]. A modification using laser illumination and optical heterodyning could improve the sensitivity of this method [4.18].

4.1.2 Connection to Processes Characterizing the Phase Transition

In the experimental investigations of phase transitions there is interest in the knowledge of certain quantities, which show a characteristic behavior close to a continuous or second-order phase transition. Such quantities are the order parameter, its static and dynamic response (the spatial and temporal distributions of fluctuations), and the specific heat capacity. Their essence is contained in the critical exponents, which are expected to help theory to describe the materials investigated by certain theoretical models.

Ultrasonic studies are relevant in this context, if the elastic response functions can somehow be related to the quantities mentioned above. This is in general the case, because there always exists some coupling between the strain and the ordering quantity. An observable indication for this fact is the spontaneous strain that appears in connection with ordering and which can be determined by dilatometry, by X-rays or by elastic neutron scattering. The differences lie not only in the magnitude but predominantly in the type of coupling.

In Sects.4.2.1-4 a general theoretical survey is given on how the elastic response is related to the quantities characterizing the phase transition and which characteristic variations in the elastic behavior are to be expected. In Sect.4.3 the experimental results for various materials are discussed in detail.

4.2 Theoretical Analysis

In this section we relate the elastic response function to the relevant interactions and processes occurring at a structural phase transition. We present this analysis in steps of increasing sophistication. We start with the Landau theory, including contributions from the strain. Symmetry considerations help to decide which terms in the expansion of the free energy density can appear and which have to be excluded. In the next subsection we leave the safe ground of mean-field theory and attempt to relate in a more general way the elastic response to the commonly used critical exponents. Some results of scaling and mode-mode coupling will be reported here. Finally the notion of marginal dimensionality is introduced and the scarce applications of the renormalization group analysis upon elastic properties are mentioned briefly.

4.2.1 Landau Theory

We start with an expansion of the free energy density in terms of the order parameter

$$F(Q,T) = F_0(T) + \frac{1}{2} a(T)Q^2 + \frac{1}{4} bQ^4 + \dots \tag{4.3}$$

with $a = a'(T - T_0)$ near T_0. The strain e gives an elastic energy contribution $\frac{1}{2} c_0 e^2$ with the background elastic constant c_0 taken at zero-order parameter. Due to the coupling of the strain to the ordering quantity an interaction energy density F_{int} has to be added, which is phenomenologically expanded in powers of e and Q

$$F_{int} = g\, e\, Q + h\, e\, Q^2 + \dots \quad . \tag{4.4}$$

To keep the exposition simple, the discussion is given here with only one component of e and Q. The general case will be treated in the next subsection.

The ordering quantity plays the role of an internal degree of freedom. It can move more or less freely under the action of forces exerted by the ultrasonic strain field and described by F_{int}. The ordering quantity responds to these forces and reacts back on the elastic system. The result is a change, in general a decrease, in the elastic stiffness. Therefore the most important information about structural phase transitions is obtained from the temperature dependence of the elastic functions.

In the case of bilinear coupling, $F_{int} = g\, e\, Q$, these forces are proportional to the strain only. There are no other forces

$$\frac{\partial F}{\partial Q} = ge + \left(\frac{\partial^2 F}{\partial Q^2}\right)_e \delta Q = ge + \frac{\delta Q}{\chi_Q} = 0 \quad , \tag{4.5}$$

and the ordering quantity can in general follow freely the applied varying strain: $\delta Q = -\chi_Q\, ge$. Its response is determined by the unrenormalized order parameter susceptibility χ_Q. The moving ordering quantity δQ in turn adds a contribution to the stress σ acting within the sound wave

$$\sigma = \frac{\partial F}{\partial e} = c_0\, e + g\delta Q = (c_0 - g^2\chi_Q)e \quad .$$

The result is a diminished elastic stiffness, which is in the static limit equal to

$$c_T = c_0 - g^2\chi_Q = c_0\, \frac{T - T_0 - g^2/a'c_0}{T - T_0} \quad . \tag{4.6}$$

In (4.6) $T_0 = T_0^e$ is, in the case of a continuous transition, the transition temperature for zero strain (clamped state), in the absence of strain interaction (4.4).

$$T^{\sigma} = T_0 + g^2/a'c_0 \tag{4.7}$$

is the transition temperature in the free state (zero stress). For discontinuous transitions T_0 denotes the lower stability limit. If this bilinear coupling prevails, the elastic stiffness probes directly the order parameter susceptibility. This is the case not only for the static or low-frequency response, but also for the general dynamic response throughout the whole frequency range.

The imaginary part of χ_Q^e is the source for the critical ultrasonic attenuation. In many cases the order parameter response can be described by a relaxation process. This results in a dispersion for the real part of the elastic constant

$$c_T(\omega) - c_T(\omega = 0) = g^2\chi_Q^e(0)\,\frac{\omega^2\tau^2}{1 + \omega^2\tau^2} \tag{4.8}$$

and an attenuation

$$\alpha(\omega) = \frac{g^2\chi_Q^e(0)\omega^2\tau}{2\rho v_s^3(1 + \omega^2\tau^2)} \quad . \tag{4.9}$$

For $\omega\tau \gg 1$, $c_T(\omega)$ approaches $c_T(\omega = 0) + g^2\chi_Q^e(0)$ which is, according to (4.6), the background elastic function c_0.

In this representation both the order parameter susceptibility and the relaxation time are taken at constant strain (χ_Q^e, τ^e). According to the Landau-Khalatnikov approximation [4.19] the relaxation time is connected with the order parameter susceptibility through a kinetic coefficient L,

$$\tau^e(T) = \frac{\chi_Q^e(T)}{L} \quad . \tag{4.10}$$

In its simplest form L is a constant. It can be calculated using mode-mode coupling. The idea behind the Ansatz (4.10) is that the rate at which the order parameter, after a perturbation, returns to the equilibrium value, is proportional to the instantaneous nonequilibrium force on Q. It describes in a simple way critical slowing down.

Summarizing the analysis of bilinear coupling of strain and order parameter, we expect an elastic instability $c_T \to 0$ at T_0^{σ} and an ultrasonic attenuation increase towards T_0^{σ}. The limit $\omega\tau = 1$ is practically not reached in ultrasonic experiments at structural transitions, but Brillouin scattering can show dispersive effects.

In the case where the strain couples with the square of the order parameter, $F_{int} = heQ^2$, there is no effect on the elasticity above T_0 except for fluctuations.

Below T_0 the developing order parameter $\langle Q\rangle$ generates a force from the strain upon Q that increases with $\langle Q\rangle$. This results in a decrease of the elastic constant [4.20]

$$c(T < T_0) = c_0 - 4h^2\langle Q\rangle^2\chi_Q(T) = c_0 - 2\,\frac{h^2}{b} \tag{4.11}$$

which is constant in the frame of Landau theory. This variation can be related to a step in the specific heat ΔC,

$$c(T) = c_0 - 4h^2\Delta C(T)/a'^2T \quad . \tag{4.12}$$

Assuming a single relaxation time for $\langle Q\rangle$, formulae similar to (4.8,9) can be written down for the dynamic case [4.21] and for $T < T_0$. For $T > T_0$ the strain couples to order parameter fluctuations only and more sophisticated transport theories, such as mode-mode coupling have to be applied.

If one expands F_{int} into higher orders, one gets anharmonic terms of the form

$$F_{int} = \kappa e^2 Q + \lambda e^2 Q^2 \quad . \tag{4.13}$$

The corresponding contributions to the elastic constants in the ordered phase are proportional to $\langle Q\rangle$ and $\langle Q\rangle^2$. In favorable cases where domain effects are absent, a measurement of such elastic constants can give the temperature dependence of the order parameter.

A type of coupling, connecting the strain with the spatial derivative of the ordering quantity, has recently been proposed [4.22].

$$F_{int} = \nu e\,\frac{\partial Q}{\partial x} \quad .$$

This has, however, not been found experimentally yet.

4.2.2 Symmetry Aspects

The ordering quantity forms a basis for an irreducible representation of the space group of the crystal in the disordered phase. It can have up to n = 3 components Q_i for ferrodistortive transitions, but can have more in antiferrodistortive ones. This representation subduces in the ordered phase the identity representation A_{1g}. In the most general case Q_i are spatial Fourier components with wave vector q_0. The special case $q_0 = 0$ describes a ferrodistortive transition. In an antiferrodistortive transition a superstructure appears and q_0 is located at the boundary of the Brillouin zone. Cases where q_0 is in the interior of the Brillouin zone but not at $q_0 = 0$ are met for helical arrangements, for charge density waves, and generally for incommensurate phases.

In order to handle strain-order parameter coupling it is advantageous to use a combination of strain components that are adapted to the symmetry of the crystal, i.e., that form bases for irreducible representations of the space group [4.23].

Table 4.1.

Symmetry strain	Symmetry elastic const.	Representation of O_h
$e_a = e_1 + e_2 + e_3$	$c_a = \frac{1}{3}(c_{11} + 2c_{12})$	A_{1g}, Γ_1^+
$e_{e1} = e_1 - e_2$; $e_{e2} = (2e_3 - e_1 - e_2)/\sqrt{3}$	$c_e = \frac{1}{2}(c_{11} - c_{12})$	E_g, Γ_3^+
$e_t = \begin{cases} e_4 \\ e_5 \\ e_6 \end{cases}$	$c_t = c_{44}$	T_{2g}, Γ_5^+

As the wave vector of the ultrasonic wave, even at the highest frequencies used, is very small compared to the size of the Brillouin zone, the representations for the ultrasonic strain field can always be taken in the limit $\underline{q} = 0$.

In the section on experimental results, we shall deal in the majority of cases with crystals which have cubic symmetry in their high symmetry phase. We therefore give in Table 4.1 symmetrized strains and corresponding symmetrized elastic constants for cubic symmetry.

This means that the three elastic constants c_{11}, c_{12}, and c_{44} and the six components e_i of the symmetric strain tensor can be combined to the symmetrized elastic constants and symmetrized strains, which are the volume strain e_a, the tetragonal and orthorhombic strains e_{e2}, e_{e1}, and the trigonal strains e_4, e_5, e_6.

The condition for the existence of a certain type of coupling is that both partners in the coupling terms of (4.4) or (4.13) must appear with the same irreducible representation. For bilinear coupling ($g \neq 0$) strain and order parameter components must have at least one irreducible representation in common. For higher order coupling the corresponding symmetric squares have to be compared. From this it follows immediately that for a nonferrodistortive transition with $\underline{q}_0 \neq 0$ the strain can only couple with even powers of the ordering quantity, except for cases where one has a set of different $\underline{q}_0$ vectors (star of $\underline{q}_0$).

In the case of cubic symmetry, the Landau expansion (4.3,4) can also have cubic terms in the strain components. These third-order invariants are determined by the symmetry of the irreducible representation of the elastic constant matrix and are given for the different strains as follows [4.24]:

$$F_3 = c_{aaa}\, e_a^3 + c_{aee}\, e_a(e_{e1}^2 + e_{e2}^2) + c_{att}\, e_a(e_4^2 + e_5^2 + e_6^2)$$
$$+ c_{eee}\, e_{e2}(e_{e2}^2 - 3e_{e1}^2) + c_{ett}\left[e_{e1}(e_4^2 - e_5^2) + \frac{e_{e2}}{\sqrt{3}}(2e_6^2 - e_4^2 - e_5^2)\right] + c_{ttt}\, e_4\, e_5\, e_6 \quad . \tag{4.14}$$

If the strain is the order parameter the phase transition is of first order, although it can be practically of second order if the coupling constants $c_{ij\ell}$ are sufficiently small.

A more subtle point is the occurrence of a first-order phase transition as predicted by symmetry considerations and the renormalization group equations in cases where the Landau theory would still give a second-order transition. It has been shown that if the number of components of the order parameter n is bigger than four (n > 4), the ε expansion does not yield a stable fixed point and hence the transition is of first order [4.25]. The examples discussed so far in this way are magnetic Hamiltonians. There is one case of a structural transition with n = 4 (NbO_2) which can have a second-order transition [4.26]. Other examples of structural transitions have not been discussed yet.

4.2.3 Mode-Mode Coupling and Scaling

The Landau theory in its simplest form neglects completely the fluctuations of the ordering quantity $\delta Q_i = Q_i(t) - \langle Q \rangle$. These fluctuations can give, however, important contributions to c_{ij} and α. Theoretical treatments of transport coefficients incorporating these fluctuations are known as mode-mode coupling theories [4.27]. The physical picture behind it is that there is a whole spectrum of fluctuations $\delta Q_i(\underline{q})$ with different wave vectors $\underline{q}$, centered around $\underline{q}_0$, the wave vector of the order parameter. In displacive phase transitions, for example, these fluctuations are formed by the soft part of the optical phonon branch. In order-disorder transitions these modes belong to collective tunnelling motions.

Assuming a coupling of the form $F_{int} = h_{ijm} Q_i Q_j e_m$ (4.4) two fluctuations δQ of nearly opposite wave vectors combine and produce by anharmonic interaction a stress fluctuation of nearly zero wave vector. The total stress is the sum over the whole fluctuation spectrum.

$$\delta\sigma_m = \sum_{\underline{q},i,j} h_{ijm} \delta Q_i(\underline{q}) \delta Q_j(-\underline{q}) \quad . \tag{4.15}$$

A straightforward way to calculate the elastic response is to write these quantities as correlation functions (fluctuation-dissipation theorem and equipartition theorem) [4.28].

$$c_m(T) = c_m^0 - \frac{V}{k_B T} < |\delta\sigma_m|^2 > \quad . \tag{4.16}$$

$$\alpha_m(T) = \frac{\omega^2 V}{2\rho v_s^3 k_B T} \int_0^\infty e^{i\omega t} < \delta\sigma_m(t) \delta\sigma_m^*(0) > dt \quad . \tag{4.17}$$

Equation (4.17) means that, as in Brownian motion, the sound attenuation coefficient can be expressed as the Fourier transform of time-dependent correlation functions of random stresses originating from (4.4). Inserting (4.15) leads to static and dynamic four-point correlation functions of the ordering quantity of the form

$$\langle ijk\ell\rangle = \langle \delta Q_i(q,t)\delta Q_j(-q,t)\delta Q_k(q',0)\delta Q_\ell(-q',0)\rangle \quad .$$

Their temperature dependence is different according to whether equal or different components $ijk\ell$ are involved. This leads to different critical behavior for different symmetrized strain components [4.29,30]. For example, in cubic crystals c_a (bulk modulus) involves Tr$\langle iijj\rangle$, from which one can project a term describing energy-density fluctuations. For this term the elastic stiffness constant is represented, apart from numerical factors of order unity, by (4.11). It diverges with the specific heat exponent α

$$T[c_a^0 - c_a(T)] = h^2 \frac{\Delta C(T)}{a'^2} = \left(\frac{h}{a'}\right)^2 \left[A\left(\frac{T}{T_0} - 1\right)^{-\alpha} + B\right] \quad . \tag{4.18}$$

The amplitudes A and B are in general different above and below T_0. The other elastic constants c_e and c_t and their corresponding attenuation depend on correlation functions with mixed indices. Their temperature variation is in general not governed by the specific heat and will be briefly discussed in the next section.

The attenuation involves in addition the dynamics of the system. Theories developed in the past were mainly applied to the liquid-gas and magnetic phase transitions [4.31,32]. One publication deals with the structural phase transition in $SrTiO_3$ [4.30]. The additional parameter is the dynamical critical exponent z, describing the temperature and wave vector dependence of the characteristic decay time of δQ fluctuations [4.33].

$$\tau(q,\varkappa) = \varkappa^z f(q/\varkappa) \sim \left|\frac{T}{T_0} - 1\right|^{-\nu z} \quad \text{for} \quad q \ll \varkappa \quad . \tag{4.19}$$

Here $\varkappa$ denotes the inverse correlation length and f is a homogeneous function in q and $\varkappa$ and is finite in the limit $q/\varkappa \to 0$. The attenuation belonging to c_a is expected to vary with a critical exponent

$$\rho = \alpha + \nu z = 2 - \nu(d - z) \quad . \tag{4.20}$$

The theoretical problem is to relate z to static critical exponents. In one case, $SrTiO_3$, one gets $z = 2 - \eta$ from a mode-mode coupling treatment. These findings will be compared with a more recent calculation in the next section and contact will be made with experiments in the experimental sections.

4.2.4 Renormalization Group Analysis

The success of the renormalization group approach to critical phenomena was to distinguish between relevant and irrelevant parameters and to give a formal procedure to calculate certain critical exponents for static and dynamic problems [4.34,35].

As an example for the distinction of relevant and irrelevant parameters let us consider the case of $SrTiO_3$ under uniaxial stress, which was studied theoretically [4.36]. It was found that uniaxial stress acts as a modifying or symmetry-breaking field. In the unstressed crystal the order parameter (oxygen-octahedra rotation angle) has three choices to orient along one of the cubic axis. This is the case of an order parameter with three components ($n = 3$), and the system is described by a three-dimensional Heisenberg Hamiltonian. For sufficiently large uniaxial tension along a [100] direction the order parameter is directed along the stress axis. In this case the appropriate description is the Ising model with $n = 1$. On the other hand, uniaxial compression forces the order parameter into the plane perpendicular to the stress axis and the number of order parameter components becomes $n = 2$. The stress-free state itself is a bicritical point. This has important consequences on the temperature variation of elastic constants. From the free enthalpy density, in a scale invariant form,

$$G(\varepsilon,\sigma_e) = \varepsilon^{2-\alpha}F(\sigma_e/\varepsilon^{\phi_e}) \quad \text{with} \quad \varepsilon = T/T_0 - 1 \quad , \tag{4.21}$$

one can see that the symmetry-breaking stress σ_e is a relevant parameter as long as the axial crossover exponent ϕ_e is positive. Its response function, the elastic compliance

$$s_e = \left(\frac{\partial^2 G}{\partial \sigma_e^2}\right)_T \sim \varepsilon^{2-\alpha-2\phi_e} = \varepsilon^{-\mu_e} \tag{4.22}$$

varies, as a function of reduced temperature ε, with a critical exponent $\mu_e = 2\phi_e + \alpha - 2$. This is markedly different from the specific heat exponent α valid in cases where the applied ultrasonic stress is not a symmetry-breaking field (see Sect.4.2.3).

Quite recently the renormalization group analysis has been applied also to dynamical phenomena. Critical exponents for ultrasonic attenuation have been calculated [4.29] by an expansion in n and 4-d. The important difference between the behavior of Im{c} for different representations Γ for the symmetrized strains e_Γ has been stressed. For a cubic crystal, described by an isotropic Heisenberg model ($d = n = 3$), one obtains, for example,

$$\rho_a = 2 - \nu(d - 2 - c\eta) = 1.34$$

$$\rho_t = 2\phi_t - \nu(d - 2 - c\eta) = 1.86 \quad \text{with} \quad c = 6 \ln\frac{4}{3} - 1$$

$$\rho_e = 2\phi_e - \nu(d - 2 - c\eta) = 1.86 \quad . \tag{4.23}$$

Again the crossover exponents ϕ_t and ϕ_e, characterizing the combinations Q_iQ_j and $Q_i^2 - Q_j^2$, appear in the formulae and demonstrate the difference between the various modes.

This also has important consequences on the experimental determination of critical exponents. In general, the sound velocities and the corresponding attenuations depend on a sum of contributions from symmetrized components. The longitudinal wave along the cubic axis of a cubic crystal depends, for example, on c_a and c_e,

$$c_{11} = c_a + \frac{4}{3} c_e \quad . \tag{4.24}$$

We obtain a superposition of two different critical attenuations and, depending on the ratio of their amplitudes, a crossover can occur. These aspects, together with other experimental complications, will be discussed in Sect.4.3.2.

Another important notion to be mentioned in this context is that of marginal or critical dimensionality d^*[4.37-40]. For $d > d^*$ the fluctuations play a minor role and the behavior can well be described by the mean-field approximation. For $d < d^*$ we get a fluctuation-dominated behavior, which is often called "nonclassical". In the case $d = d^*$ the mean-field results have to be taken with some weakly divergent logarithmic corrections. In an isotropic system with short-range interactions d^* is usually equal to 4. Here the region of critical fluctuations scales like the correlation length ξ in all d directions. This is no longer the case if long-range dipole interaction prevails. There the correlation range increases faster along the dipole axis (proportional to ξ^2) and d^* becomes 3. Important cases are structural phase transitions with soft acoustic modes [4.39,40]. In such materials the softening is sometimes limited to certain planes or certain directions. Examples are the softening of modes depending on $c_t = c_{44}$ and $c_e = (c_{11} - c_{12})/2$ in cubic crystals. In the first example the propagation direction is limited to the three (100) planes and the polarization is always perpendicular to these planes. In the second example the soft modes are confined to the six [110] directions with their wave vector and to [1$\bar{1}$0] with their polarization. Consequently the correlated regions in direct space grow as for dipolar interaction as $\xi \times \xi \times \xi^2$ in the first case and proportional to $\xi^2 \times \xi^2 \times \xi$ in the second. The resulting marginal dimensionalities are $d^* = 3$ and 2, respectively. For the softening of the bulk modulus c_a the marginal dimensionality $d^* = 0$. Applications of the various cases mentioned here to real systems will be given in the result section.

4.3 Experimental Results

In the following sections experimental results of elastic properties near structural phase transitions are presented. The main emphasis is put on the real part of the elastic functions, since here more information concerning the phase transition and its mechanism can be obtained. The attenuation, as a transport coefficient, is in general more complicated.

The various materials are arranged with respect to their order parameters. The main distinction is whether a component of the strain tensor can play the role of the order parameter or not. For each class of materials we shall pick out one or two compounds for which the study of elastic properties played an essential role in elucidating the mechanism leading to the structural change. All the other materials we shall briefly summarize at the end of each section. Clearly this selection is not always unique and often biased, but we tried to choose examples where the ultrasonic method clearly gave as much or more information than other techniques such as inelastic neutron scattering, optical spectroscopy, etc.

4.3.1 Strain Is Order Parameter

This is an important class of materials for two reasons. Firstly, the elastic response function, measured in the ultrasonic experiment, shows as the response of the ordering quantity, the strongest variation at the phase transition. We observe a softening of a certain symmetric combination of elastic stiffness functions, an elastic instability. In the phonon picture there are soft acoustic modes with velocities depending on these combinations. Secondly, as the softening of acoustic phonons is often limited to certain planes or directions, the correlations of critical fluctuations become rather anisotropic and marginal dimensionalities $d^* = 3$ or 2 result. This means that in three dimensions either pure mean-field behavior is to be expected ($d^* = 0$ or 2), or mean-field behavior with logarithmic corrections ($d^* = 3$), as discussed in Sect.4.2.4. The fact that the strain acts as order parameter also implies cases where other physical quantities exist that are proportional to the strain in question, such as electric polarization, a phonon or electron coordinate. These other quantities can equally serve the same purpose. The essential point is that there is a bilinear interaction between strain and these quantities, as indicated by $g_{im} \neq 0$ in (4.4). We shall call such phase transitions "piezodistortive", to indicate their analogy to the piezoelectric effect. Often the term "ferroelastic" [4.41] is used, sometimes with the restriction that an elastic hysteresis loop exists and that the spontaneous strain can be reversed by a stress larger than the coercive force.

Further separation is made with respect to the quantities bilinearly coupled to the strain and often characterizing the mechanisms leading to the phase transitions: ferroelectrics, cooperative and band Jahn-Teller transitions, valence fluctuations, and others not fitting into this scheme. Physical properties of the substances can be found also in Tables 4.2-6. In the references listed in these tables one can find important material constants.

Table 4.2. Materials for which strain is order parameter: ferroelectric and nonelectronic transitions

Material	T_c [K]	Crystal classes	Spontaneous strain	Symmetry	Soft c_{mn}	Order of transition	References
Ferroelectric							
KH_2PO_4	122	D_{2d}-C_{2v}	e_6	B_2	c_{66}	1	[4.45-48]
RbH_2PO_4	145	D_{2d}-C_{2v}	e_6	B_2	c_{66}	1	[4.52]
CsH_2AsO_4	149	D_{2d}-C_{2v}	e_6	B_2	c_{66}	1	[4.54]
K-Na-tartrate	297	D_2-C_2	e_4	B_3	c_{44}	2	[4.56,57]
	255	C_2-D_2				2	
Li-NH_4-tartrate	98	D_2-C_2	e_5	B_2	c_{55}		[4.199]
$KH_3(SeO_3)_2$	211	D_{2h}-C_{2h}	e_4	B_{3g}	c_{44}		[4.200]
Nonelectronic							
KCN	168	O_h-D_{2h}	e_t	T_{2g}	c_{44}	1	[4.60-63]
	83	D_{2h}-D_{2h}					
NaCN	284	O_h-D_{2h}	e_t	T_{2g}	c_{44}	1	[4.201,202]
TeO_2	p_c=8.86 kbar	D_4-D_2	$2e_3-e_1-e_2$	B_{1g}	$c_{11}-c_{12}$	2	[4.64,65]

Table 4.3. Materials for which strain is order parameter: electronic transitions

Material	T_c [K]	Crystal classes	Spontaneous strain	Symmetry	Soft c_{mn}	Order of transition	References
Cooperative Jahn-Teller							
$Ni_xZn_{1-x}Cr_2O_4$	82 - 300 (x=0,67-1)	O_h^7-D_{4h}	e_{e2}	E_g	$c_{11}-c_{12}$	1	[4.87]
$CsCuCl_3$	423	D_{6h}-D_6	e_4	$E_1(q_0 = \frac{1}{3} c^*)$	c_{44}	1	[4.89,90]
$DyVO_4$	13	D_{4h}-D_{2h}	e_1-e_2	B_{1g}	$c_{11}-c_{12}$	2	[4.74,92]
$TbVO_4$	34.7	D_{4h}-D_{2h}	e_6	B_{2g}	c_{66}	2	[4.74,92]
TmCd	3.16	O_h-	e_{e2}	E_g	$c_{11}-c_{12}$	1	[4.97]
TmZn	8.5	O_h-	e_{e2}	E_g	$c_{11}-c_{12}$	1	[4.98]
$PrAlO_3$	151	C_{2v}-C_5	e_1-e_2	B_1	$c_{11}-c_{12}$	2	[4.99]
$PrCu_2$	7.3	C_{2v}-	e_5		c_{55}	2	[4.100]
UO_2	65	O_h-	e_t	T_{2g}	c_{44}	1	[4.101]
Fe_2TiO_4	142	O_h-	e_{e2}	E_g	$c_{11}-c_{12}$		[4.91]
Fe_3O_4	120		e_t	T_{2g}	c_{44}	1	[4.2]
Band Jahn-Teller							
Nb_3Sn	45	O_h-D_{4h}	e_{e2}	E_g	$c_{11}-c_{12}$	1	[4.112]
V_3Si	21	O_h-D_{4h}	e_{e2}	E_g	$c_{11}-c_{12}$	1	[4.110]
La Ag_x In_{1-x}		O_h-D_{4h}	e_{e2}	E_g	$c_{11}-c_{12}$	1	[4.123]

Table 4.3 (continued)

Valence transition							
$Sm_{1-x}Y_xS$	(x > 0.15)	O_h	e_a	A_{1g}	$c_{11}+2c_{12}$		[4.125]
$Ce_{1-x}Th_x$	148 (x=0.272)	O_h	e_a	A_{1g}	$c_{11}+2c_{12}$		[4.126]
TmSe		O_h	e_a	A_{1g}	$c_{11}+2c_{12}$		[4.127,128]
Martensitic transition							
In_xTl_{1-x}	195 (x=0.75)	O_h-D_{4h}	e_{e2}	E_g	$c_{11}-c_{12}$		[4.129]
$Au_xCu_{y-x}Zn_{1-y}$		O_h-D_{2h}	e_{e1}	E_g	$c_{11}-c_{12}$		[4.130]
La_3Se_4					$c_{11}-c_{12}$		[4.203]
$AuCuZn_2$	60	O_h-D_{4h}	e_{e2}	E_g	$c_{11}-c_{12}$	1	[4.131]

Table 4.4. Materials for which strain is not order parameter: phonon transitions

Material	T_c [K]	Crystal classes	Order parameter	Symmetry	q_0	C_{mn} affected by transition	Order of transition	References
$SrTiO_3$	105	O_h-D_{4h}	octahedra rotation	R_4^+ (=T_{1g})	$\frac{\pi}{a}$ (1,1,1)	$c_{11}-c_{12}$, c_{44}, $c_{11}+2c_{12}$	2	[4.133-136]
$KMnF_3$	187	O_h-D_{4h}	octahedra rotation	R_4^+ (=T_{1g})	$\frac{\pi}{a}$ (1,1,1)	$c_{11}-c_{12}$, c_{44}· $c_{11}+2c_{12}$	1	[4.142]
	91.5	D_{4h}-D_{4h}	octahedra rotation	M_3	$\frac{\pi}{a}$ (1,1,0)		1	[4.143]
$CsPbBr_3$	403	O_h-D_{4h}	octahedra rotation	M_3	$\frac{\pi}{a}$ (1,1,0)	$c_{11}-c_{12}$		

Table 4.4 (continued)

	361	$D_{4h}-D_{2h}$	octahedra rotation	$\approx R_4^+$	$\frac{1}{2}(a^*,a^*,c^*)$			
$Sn_xGe_{1-x}Te$	~ 700 (x=0) ~ 100 (x=1)	O_h-C_{3v}	sublattice shift in [111]	T_{1u}	0	$c_{11}-c_{12}, c_{44}, c_{11}+2c_{12}$	1 (x < 0.73) 2 (x < 0.73)	[4.148,149]
SiO_2	846	D_6-D_3	~ tetrahedron rotation	B_1	0	$c_{11}+c_{12}, c_{33}, c_{13}; c_{14}; c_{66}$	1	[4.204]
$Gd_2(MoO_4)_3$	432	$D_{2d}-C_{2v}$		M_1+M_5	$\frac{\pi}{a}(1,1,0)$	$c_{11}+c_{12}, c_{66}$	1	[4.156,205]
$Tb_2(MoO_4)_3$		$D_{2d}-C_{2v}$		M_1+M_5	$\frac{\pi}{a}(1,1,0)$		1	[4.157,158]
$BnMnF_4$	255	$C_{4v}-C_2$			$(0.392a^*, 0.5b^*, 0.5c^*)$	$c_{11}, c_{22}, c_{33}, c_{44}, c_{66}$		[4.159,160]

Table 4.5. Materials for which strain is not order parameter: ferroelectric transitions

Material	T_c [K]	Crystal classes	Order parameter	Symmetry	q_0	c_{mn} affected by transition	Order of transition	References
$Pb_5Ge_3O_{11}$	450	$C_{3h}-C_3$	el. polarization P_3	A"	0	$c_{11}+c_{12}, c_{33}, c_{13}$	1	[4.178,179]
$BaTiO_3$	401	O_h-C_{4v}	el. polarization (P_1,P_2,P_3)	T_{1h}	0	$c_{11}-c_{12}, c_{44}, c_{11}+2c_{12}$	1	[4.175]
	278	$C_{4v}-C_{2v}$	P_1+P_2	E_u	0			

Table 4.5 (continued)

	223	C_{2v}-C_{3v}	$P_1+P_2+P_3$	B_1	0			
Ca_2Sr-propionate	282	D_4-C_4	el. polarization P_3	A_2	0	$c_{11}+c_{12}, c_{33}, c_{13}$	2	[4.171,177]
Ca_2Ba-propionate	267	O-					1	[4.206]
	203						2	
Ca_2Pb-propionate	333	D_{4h}-D_4		A_{1u}		$c_{11}+c_{12}, c_{33}, c_{13}$	2	[4.172]
	192	D_4-C_4						
$NaNO_2$	438.2	D_{2h}-			incommens.		2	[4.169,170]
	437	-D_{2v}	el. polarization P_2	B_{2u}	0	c_{11}, c_{22}, c_{33}	1	
Cu (H COO)$_2$.4 H_2O	234	C_{2h}-	antiferroel. polarization P_2	B_u	$\neq 0$	$c_{33}, c_{11}, c_{22}, c_{12}, c_{13}, c_{23}, c_{55}, c_{15}, c_{25}, c_{35}$	1	[4.207]
$(NH_2\ CH_2\ COOH)_3$ H_2SO_4 (TGS)	322	C_{2h}-C_2	el. polarization P_2	A_u	0	$c_{11}, c_{22}, c_{33}, c_{66}$ c_{12}, c_{13}, c_{23}	2	[4.164]
$(NH_2\ CH_2\ COOH)_3$ H_2SeO_4	296	C_{2h}-C_2	el. polarization P_2	A_u	0	$c_{11}, c_{22}, c_{33}, c_{66}$ c_{12}, c_{13}, c_{23}	2	[4.166]
SbSI	292	D_{2h}-C_{2v}	el. polarization P_3	B_{1u}	0	$c_{11}, c_{22}, c_{33}, c_{13}$ c_{12}, c_{23}, c_{66}	1	[4.173]

Table 4.6. Materials for which strain is not order parameter: orientational, charge density waves, solid electrolytes and incommensurate phases

Material	T_c [K]	Crystal classes	Order parameter	Symmetry	q_0	c_{mn} affected by transition	Order of transition	References
Orientational								
NH_4Cl	243	O_h-T_d	NH_4-order (octupole)	A_{2u}	0	$c_{11}+2c_{12}$	1	[4.181]
NH_4Br	234	O_h-D_{4h}	NH_4-order (octupole)	$M_3^-(\hat{=}B_{1u})$	$\frac{\pi}{a}$ (1,1,0)	$c_{11}-c_{12})$	1	[4.186]
Charge density waves								
TTF-TCNQ	52	C_{2h}	Peierls distortion,		$\frac{a^*}{2}(1-\delta)$, 0.295b^*; 0)	c_{22}, c_{66}		[4.189]
$K_2Pt(CN)_4Br_{0.3}$ 3.2 H_2O		C_{4v}	charge-density wave amplitude gap		$\frac{a^*}{2}, \frac{a^*}{2}, 0.3c^*$	c_{44}, c_{66} ...		[4.190]
2H-$TaSe_2$	122 92.5	D_{6h}	charge-density wave amplitude gap		$\frac{a^*}{2}(1-\delta)$	c_{11}...	2 1	[4.192]
2H-$NbSe_2$	29.8	D_{6h}	charge-density wave amplitude gap			c_{11}...	2	[4.192]
Solid electrolytes								
$RbAg_4I_5$	208 122	O-D_3 D_3-D_3				$c_{11}+2c_{12}, c_{44}$, $c_{11}-c_{12}$	2 1	[4.197,198]

Table 4.6 (continued)

Incommensurate phases							
K_2SeO_4	127	D_{2h}-inc.-	Σ_2	$\frac{a^*}{3}(1-\delta)0,0$	$c_{33}, c_{44},$	2	[4.209-211]
	93	C_{2v}			c_{55}	1	
Rb_2ZnCl_4	302	D_{2h}-inc.-	Σ_2	$\frac{a^*}{3}(1-\delta)0,0$	c_{33}, c_{22}, c_{44}	2	[4.212]
	192	C_{2v}					
$[N(CH_3)_4]_2ZnCl_4$	293	D_{2h}-inc.-	Σ_2	$\frac{a^*}{3}(1-\delta)0,0$	c_{11}, c_{22}, c_{33}	2	[4.213]
	280	C_{2v}					
$(NH_4)_2BeF_4$	183	D_{2h}-inc.-	Σ_2	$\frac{a^*}{3}(1-\delta)0,0$	c_{22}, c_{55}		[4.214]
	177	C_{2v}					
$K_2PbCu(NO_2)_6$	281	T_h-inc.-		$(0.42, 0.42, 0)\ a^*$	$c_{33};\ c_{11}-c_{12}$		[4.215]
	173						

a) *Ferroelectrics*

Crystals discussed in this subsection develop both a spontaneous strain and a spontaneous electric polarization in the ordered phase. Piezoelectric, i.e., bilinear, coupling connects $\underline{P}$ and $\underline{\underline{e}}$. We shall discuss potassium dihydrogen phosphate (KDP) in detail because ultrasonic propagation studies give a great deal of information about the phase transition.

Potassium Dihydrogen Phosphate (KH_2PO_4)

Its phase transition is triggered by the ordering of protons within the hydrogen bonds connecting adjacent PO_4 tetrahedra. Linearly coupled with this process are atomic displacements of K and PO_4 along the c axis, producing the electric polarization P_3 and the shear strain e_6 in the a-b plane [4.42]. All three modes belong to the same irreducible representation B_2. The proton ordering process in the double-well potential of the hydrogen bond is well suited for a description by the pseudospin formalism with $s^z = \pm 1$ [4.43]. The Hamiltonian contains contributions from atomic displacements (optical phonons), strain (acoustic phonons), the tunneling splitting Ω, and terms for the coupling between pseudospin, strain, and atomic displacements. By a transformation to what is called "displaced phonon coordinates" the optic phonon-pseudospin interaction can be written in the form of an effective exchange interaction between pseudospins

$$H = -\frac{1}{2}\sum_{ij}{}' J(i,j)s^z(i)s^z(j) - \sum_i [ge_6 s^z(i) + \Omega s^x(i)] \quad . \tag{4.25}$$

The summations are taken over all lattice sites i and j. This equation is formally identical to a magnetic Ising spin system in external fields along the x and z axes. It is also used in the description of the cooperative Jahn-Teller effect discussed below.

Already in 1946 MASON noticed the difference between c_{66}^P and c_{66}^E, the shear stiffnesses measured in an open and a short-circuited sample [4.44]. Whereas c_{66}^P behaves normally, c_{66}^E decreases drastically when the transition at about 122 K is approached from above or below. In recent years c_{66}^E has been studied in detail both in protonated [4.45,46] and deuterated crystals [4.47], combined with measurements of the ultrasonic attenuation and Brillouin scattering of light [4.48]. Applying an electric field renders of the ferroelectric phase monodomain. Above T_c the measured shear stiffness can be well described by (4.6), based on the Curie-Weiss law for the electric susceptibility. Below T_c an eight-order term in the expansion of the free energy in terms of the polarization is needed for a proper fit, or the Silsbee-Uehling-Schmidt model [4.49]. A very satisfactory description below T_c is obtained by the pseudospin model, taking also into account the difference between adiabatic (actually measured) and isothermal (calculated directly from the free energy) stiffness.

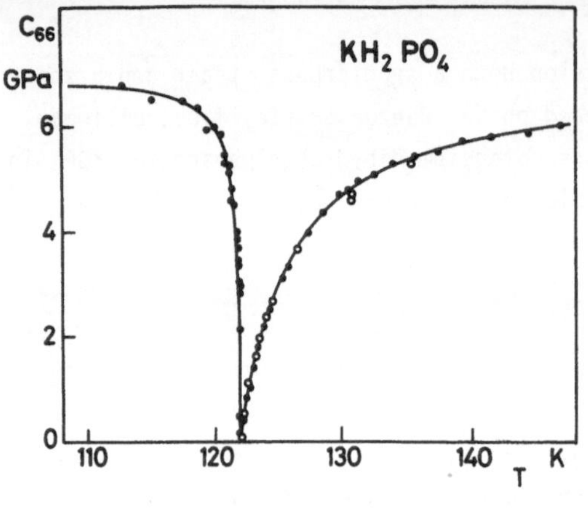

Fig.4.2. Shear stiffness c_{66} of KH_2PO_4 measured ultrasonically (o[4.45]), and by Brillouin scattering (•[4.48]). The solid line is a fit with the pseudospin model [4.43]. (Note that 1 GPa = 10^{10} dyn/cm^2)

Figure 4.2 shows the results of ultrasonic and Brillouin scattering measurements and the calculated curve from the pseudospin model [4.43].

$$c_T = c_0 \left(\frac{\Omega - J'\langle s^x \rangle}{\Omega - (J' - \frac{g^2}{c_0})\langle s^x \rangle} \right) .$$

This formula has the same form as (4.6) with $\langle s^x \rangle = \mathrm{tgh}(\Omega/kT)$ which becomes Ω/kT for small arguments. J' is the $q = 0$ Fourier transform of J_{ij} in (4.25).

The fact that Brillouin scattering gave within experimental accuracy results identical to those from ultrasonic experiments indicates that both measurements were performed at frequencies well below the relaxation rate $1/\tau$ of polarization fluctuations.

The observed ultrasonic attenuation [4.45,46] follows more or less (4.7,8). The kinetic coefficient L in the Landau-Khalatnikov Ansatz turns out to be temperature dependent, in fact more in KH_2PO_4 than in KD_2PO_4. This difference probably comes from the differing proton and deuteron tunneling frequencies. A microscopic theory for the ultrasonic attenuation [4.47,49] makes use of the so-called Takagi groups, PO_4 tetrahedra with one, three, or even four protons attached to them. These states are higher in energy than the ground state with just two protons. This is the basis of the Silsbee-Uehling-Schmidt model mentioned above. Rate equations between the ground state and various Takagi groups lead to a microscopic picture for the kinetic coefficient L. Another theoretical treatment [4.50] of the ultrasonic attenuation uses the pseudospin formalism in connection with a Green's function approach and arrives at a relationship

$$\alpha_s(T) = c_1 \omega^2 T \left(\frac{T - T_0^e}{T - T_0^\sigma} \right)^2 \qquad (4.26)$$

which describes the experimental results in protonated crystals [4.45] fairly well.

The velocity of longitudinal waves along the ferroelectric c axis and their attenuation were measured in KH_2PO_4 [4.51] and RbH_2PO_4 [4.52]. These waves couple with the pseudospin via the distance dependence of the exchange-interaction parameter J(i,j) [4.53]. Such a coupling is formally identical to the strictive terms in (4.4) containing the coefficients h. There is no elastic instability; the characteristic times derived from the attenuation therefore are quite different from the relaxation times obtained from the shear-wave attenuation [4.45]. Theory [4.53] predicts a temperature dependence different from (4.26) in the paraelectric phase. Unfortunately the attenuation measurements by both authors [4.51,52] were performed in the ferroelectric phase and do not allow any comparison.

Other Systems

The isomorphous compound *cesium dihydrogen arsenate* (CsH_2AsO_4) shows the same behavior for c_{66}^E [4.54]. Its corresponding values are, however, considerably smaller than in KH_2PO_4, a fact leading to a much larger difference between the Curie-Weiss temperature T_0^σ for zero stress and T_0^e for zero strain.

In *Rochelle salt* (potassium sodium tartrate tetrahydrate) ($KNaC_4H_4O_6 \cdot 4H_2O$) there is a piezoelectric coupling between the electric polarization P_1 and the shear strain e_4 causing the stiffness c_{44} to soften [4.55,56]. Direct measurements of this mode mode were not able to follow c_{44} to low values, so longitudinal waves under 45° against the b and c axis were used, and c_{44} was determined by subtracting the "normal" contributions c_{22},c_{33}, and c_{23}. The evaluation of the attenuation, following the Landau-Khalatnikov Ansatz, gave for the polarization relaxation time in seconds $\tau = 8.5 \cdot 10^{-11}/(T - T_0)$. All these measurements were performed at the upper transition temperature of 297 K. There is another transition at 255 K, at which ferroelectricity disappears again.

Table 4.2 lists the piezodistortive ferroelectrics that have been studied ultrasonically and gives their important parameters and references to the literature.

b) *Other Nonelectronic Compounds*

Potassium Cyanide (KCN)

In the alkali cyanides the CN group can order with respect to its orientation. In potassium cyanide two phase transitions are observed, one first-order transition at 168 K with ordering of the CN groups along the cubic [110] directions, which does not distinguish between C and N ends, and another one at 83 K, which orders with respect to head and tail [4.58]. A microscopic theory for static and dynamical properties has been set up recently [4.58,59]. An acoustic soft mode depending on c_{44} was found by ultrasonic [4.60,61] and Brillouin scattering [4.61-63] experiments. The elastic strains e_4, e_5, and e_6 are linearly coupled with a threefold degenerate set of distribution functions η_i for the ordering of the CNs. It is interesting to note that this strain order parameter coupling is essential for the phase transition, and that there is a competing ordering process coupling to the shear components e_{e1}, e_{e2} with distributions functions ξ_i of different symmetry (E_g) [4.58,61]. The coupling energy

$$F_{int} = g_t(\eta_1 e_4 + \eta_2 e_5 + \eta_3 e_6) + g_e(\xi_1 e_{e1} + \xi_2 e_{e2}) \qquad (4.27)$$

contains both contributions. In a simple picture η_1 denotes, for example, the difference in the relative occupation numbers of the orientations [011] and [01$\bar{1}$]. The other interaction is weaker and does not produce a phase transition but only an anomalous decrease of $c_{11} - c_{12}$ in the cubic phase. In Fig.4.3 we show the strong temperature dependence of c_{44} and $(c_{11} - c_{12})/2$ together with a fit to (4.6). A comparison between ultrasonic and Brillouin scattering data indicates that the characteristic times for the fluctuations of the ordering quantity are shorter than 10^{-10} s. The shear stiffness c_{44} was also measured in a series of mixed crystals $K(CN)_xCl_{1-x}$ [4.61]. Increasing the amount of added KCl weakens the interaction between CN groups and leads to a corresponding decrease of the transition temperatures and the stability limit of the cubic phase.

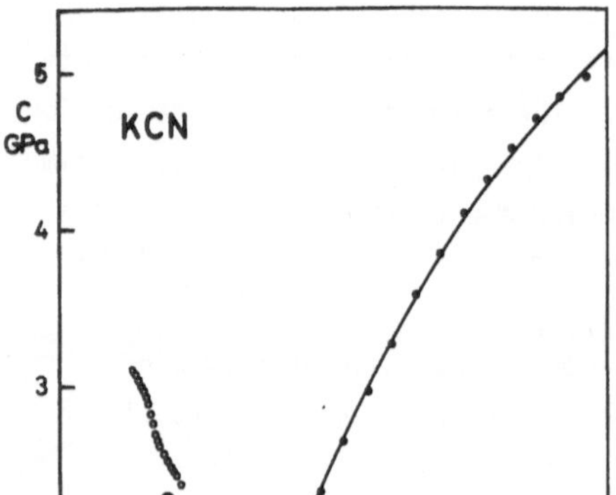

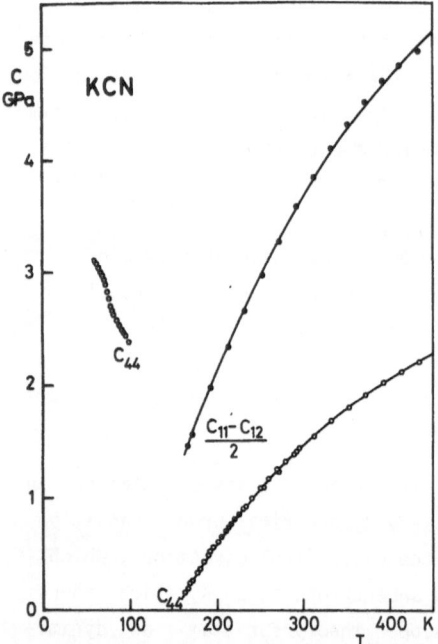

Fig.4.3. Shear stiffness $(c_{11} - c_{12})/2$ and c_{44} for KCN (•[4.60]),(○[4.61]). The solid line is a fit to (4.6) with $T_0^\sigma = 151.2$ K, $T_0^e = -230.5$ K, $c_{44}^0 = 5.24$ GPa for c_{44}, and $T_0^\sigma = 112$ K, $T_0^e = -230.5$ K, $c_e^0 = 10.4$ GPa for c_e

A material in which the phase transition is produced not by a change in temperature but by hydrostatic pressure is *paratellurite* (TeO_2). As a function of pressure the elastic function $c_{11} - c_{12}$ shows the variation characteristic for a piezodistortive transition [4.64,65]. At the transition pressure $p_c = 9$ kbar (at room temperature) an elastic instability occurs and the crystal transforms from an orthorhombic to a tetragonal structure (D_2^4 to D_4^4). Recent inelastic neutron scattering experiments [4.66] could demonstrate that the softening of the lower shear mode in [110] is limited to the innermost part of the Brillouin zone ($q < a^*/20$); the rest of the branch experiences only a shift with pressure, downward for $p < p_c$ and upward again for $p > p_c$. The microscopic mechanism for this pressure-induced phase transition is not known. Phenomenologically it can be described by proper third- and fourth-order elastic stiffness functions leading to such an extremely strong pressure dependence of $c_{11} - c_{12}$. The b/a ratio can serve as order parameter and shows mean-field behavior with $\beta = 1/2$ [4.66-68].

Electronically Driven Phase Transitions

In this group of materials the electrons play an important role and are directly involved in the transition. They are either in localized orbital states (cooperative Jahn-Teller transition), or in band states (band Jahn-Teller transitions, charge density wave transitions, and martensitic transitions in metal alloys), or in both band and localized states (valence instability transitions). In all of these cases an electronic quantity can play the role of the order parameter. In the following we shall discuss all these various transitions, where ultrasonic techniques were indispensable for characterizing the phase transition. Materials exhibiting various electronic transitions are listed in Table 4.3.

c) *Cooperative Jahn-Teller Transitions*

Here we discuss materials containing magnetic ions, i.e., ions with unfilled 3d, 4f, or 5f shells. Strong exchange interactions may give rise to magnetic phase transitions for such compounds. There is also the possibility of a structural phase transition, which is driven by the interaction between localized orbital electronic states and lattice modes. This is manifested by the splitting of degenerate electronic states and a simultaneous lowering of the crystal symmetry at a given temperature T_0, which is in general different from the magnetic transition temperatures. This mechanism, based on the single ion Jahn-Teller effect [4.69,70], was originally proposed for the structural transitions occurring in spinel structure compounds [4.71,72]. In recent years many more substances were found which exhibit structural transitions of this kind (transition metal, rare earth, and actinide compounds). There are now several review articles covering this subject [4.73-75]. We briefly outline the calculation of elastic constants and present experimental results afterwards.

Calculation of the Elastic Constants

Two models have been proposed for the cooperative Jahn-Teller effect. The first one, suitable for transition metal compounds, treats first a complex, consisting of magnetic ion plus ligand ions, and introduces afterwards a coupling between these complexes. This is called the cluster model [4.76]. In the second model, suitable for rare earth compounds, one couples the phonon states directly to the electronic states of the magnetic ions (band phonon model).

Likewise the coupling of the strain to the magnetic ion can occur in two ways [4.75]. Either the strain can couple to the local vibrational coordinates Q_Γ of a Jahn-Teller complex $H' = \sum_\Gamma g_\Gamma Q_\Gamma \varepsilon_\Gamma$, or the strain can couple directly to the electronic coordinates, the electric quadrupole moment O_Γ of the magnetic ion $H' = \sum g_\Gamma \varepsilon_\Gamma O_\Gamma$. Which of the two interactions is dominant depends on the structure and the type of magnetic ion. If the elastic deformation energy density has coupling terms between strain and internal coordinates [4.77] the first form of interaction is always present. If, however, such a coupling is absent (e.g., in structures such as NaCl and CsCl structures) only the second type of interaction is possible. For the following we choose the latter because the resulting elastic constant expressions of the two types of interaction have the same form and are difficult to distinguish [4.75].

As an effective Hamiltonian for describing the cooperative Jahn-Teller effect we take

$$H_{\text{int}} = -\sum_{i,\Gamma} g_\Gamma \varepsilon_\Gamma O_\Gamma(i) - \frac{1}{2}\sum_{ij\Gamma}' G_{ij} O_\Gamma(i) O_\Gamma(j) \quad . \tag{4.28}$$

Here the first term is the strain magnetic ion interaction discussed above, commonly called the magnetoelastic interaction term. For cubic symmetry one can have two coupling terms: g_{Γ_3} which couples the electric quadrupole operators $O_0^2 = 3J_z^2 - J(J+I)$ and $O_2^2 = J_x^2 - J_y^2$ to the corresponding symmetry strains e_{e2} and e_{e1} (Table 4.1) and g_{Γ_5} which describes the coupling $J_x J_y + J_y J_x$, etc., to e_6, etc. These terms lead to elastic anomalies for crystal field split magnetic ions. Examples are PrSb, TmSb, ..., $FeCl_2$ [4.78,79]. For a review see [4.80]. They can also lead to strong elastic softening for magnetic ions with orbitally degenerate ground states (Γ_3, Γ_4, Γ_5, and Γ_8 for ions with cubic point symmetry), which ultimately can lead to a structural transition where the elastic constant $c_\Gamma = c_0 - g_\Gamma^2 \chi_s^\Gamma(T)$ becomes zero. Here $\chi_s^\Gamma = d\langle O_\Gamma\rangle/d\varepsilon_\Gamma$ is the single ion strain susceptibility. The structural transition in this case originates in a strain field coupling between the magnetic ions, giving an effective quadrupole-quadrupole interaction of infinite range $(g^2/c_0)\sum_{ij} O_\Gamma(i) O_\Gamma(j)$ similar to the cases of KH_2PO_4 and KCN discussed above.

The second term in (4.28) describes the coupling between magnetic ions which has its origin in different mechanisms than the magnetoelastic interaction. The coupling G_{ij} can arise for example due to an interaction of magnetic ions with optical phonons, or $k \neq 0$ acoustic phonons, or conduction electrons in the case of intermetallic compounds (aspherical Coulomb charge scattering) [4.81,82]. The simple form of (4.28) is very convenient for calculating elastic constants, but unfortunately can be rigorously justified only in few cases [4.73] in the band model.

The elastic constant resulting from H_{int} of (4.28) can be derived in mean-field theory in a way analogous to (4.6) [4.73,83-85].

$$c_\Gamma/c_\Gamma^0 = \frac{1 - (g^2{}_\Gamma/c^0 + g')\chi_s}{1 - g'\chi_s} \quad . \tag{4.29}$$

Here $g' = \sum'_{ij} G_{ij}$ is the $\underset{\sim}{q} = 0$ component of the Fourier transform of the quadrupolar constant G_{ij}, with self-energy terms taken into account.

Note the similar structure of (4.29) and (4.6). This similarity becomes even more transparent if one neglects higher lying crystal field energy levels of the magnetic ion. In this case for a degenerate ground state level the strain susceptibility χ_s exhibits a 1/T temperature dependence and (4.29) has exactly the same form as (4.6). This is the case, for example, for transition metal compounds, where the higher lying crystal field energy levels are typically 10^3 cm^{-1} above the ground state. This discussion is analogous to the one given for KPD (Sect.4.31a).

Experimental Results

Ultrasonic measurements provide a great deal of information about this class of phase transitions. This is because due to time reversal symmetry the strain can couple to the electric quadrupoles, whereas with neutrons and Raman scattering one can observe vibronic excitations only via phonon hybridization or in the presence of magnetic fields. From the temperature dependence of the soft elastic constants one can determine the constants g_Γ and g' (4.29). g_Γ determines the magnitude of the spontaneous strain and g' gives an indication of other than strain coupling between the magnetic ions. The materials studied in greatest detail so far are the rare earth vanadates and arsenates with zircon structure. They have been reviewed in detail recently [4.73,74]. We therefore shall only make a brief remark about these compounds. We shall discuss one material each from transition metal, rare earth, and actinide compounds. All these substances with one exception ($CsCuCl_3$) show ferrodistortive transitions.

Transition Metal Compounds: The best studied system so far is *nickel-zinc-chromite* ($NiCr_2O_4$). The Jahn-Teller ion is the $Ni^{2+}(3d^8)$, which is located on a tetrahedral site of the normal spinel lattice. The $^3F_4$ ground state splits in a cubic field into Γ_4, Γ_5, and Γ_2 levels, where only the lowest triplet state is thermally populated in the temperature region of interest. $NiCr_2O_4$ exhibits a cubic to tetragonal phase transition at T_0 = 300 K and a magnetic phase transition at T_N = 72 K. The structural phase transition has been investigated with ultrasonic, X-ray, and thermal expansion techniques [4.86,87]. The phase diagrams of $Cu_{1-x}Ni_xCr_2O_4$ and $Fe_{1-x}Ni_xCr_2O_4$ have been studied by KATAOKA and KANAMORI [4.84] using the displaced phonon formalism leading to (4.28,29). Therefore we interpret the results accordingly, although the strain coupling to vibronic states, mentioned above, might be important and the system could be described with the cluster model as well.

In Fig.4.4 we show for $NiCr_2O_4$ the temperature dependence of the sound velocity for the soft mode $c_{11} - c_{12}$, as well as the c/a ratio, which is proportional to the order parameter. For $T > T_a$ the temperature dependence of the sound velocity was fitted with (4.29) using $\chi_s = 72/T(K^{-1})$ appropriate for the Γ_4 ground state and taking $g' \ll g^2/c_0$. This mode would become completely soft at T_0 = 274 K; however, the structural transition occurs at T_a = 300 K. In the tetragonal phase no attempt was made to fit the elastic data, because we do not deal with a single domain state here. With the parameters determined from the elastic measurements, one can also interpret the temperature dependence of the order parameter. Neglecting now the g' term, we obtain from $\partial F/\partial e_{e2} = 0$ the order parameter equation for the tetragonal strain [4.84,87],

$$e_{e2}/e_{e2}^0(T=0) = \frac{\exp(3e_{e2}T_0/e_{e2}^0 T) - 1}{\exp(3e_{e2}T_0/e_{e2}^0 T) + 2} \quad . \tag{4.30}$$

The temperature dependence of the tetragonal strain e_{e2} is fitted to the experimentally determined c/a ratio as shown in Fig.4.4. Again the agreement between theory and experiment is very good. Especially noteworthy are the following facts:

1) Equation (4.30) predicts at the transition temperature T_a, e_{e2}/e_{e2}^0 = 0.5, in agreement with the experiment. The first-order nature of the transition is due to the fact that the Ni^{2+} ion is in a triplet state.

2) Equation (4.30) predicts for the ratio $T_a/T_0 = 1.5/\ln 4 = 1.082$. This has to be compared with the experimentically determined ratio 299/274 = 1.091, again in close agreement.

3) The spontaneous strain at T = 0 neglecting distortions due to the magnetic transition at 72 K is from experiment (c - a)/a = 0.042. This can be compared with the calculated value determined from the parameters quoted above:
(c - a)/a = 0.060.

In summary the cooperative Jahn-Teller phase transition in $NiCr_2O_4$ is well understood. The main feature is an effective strain-magnetic ion coupling which dominates all other possible optical phonon coupling. This feature has also been verified with inelastic neutron scattering and diffuse X-ray scattering [4.88].

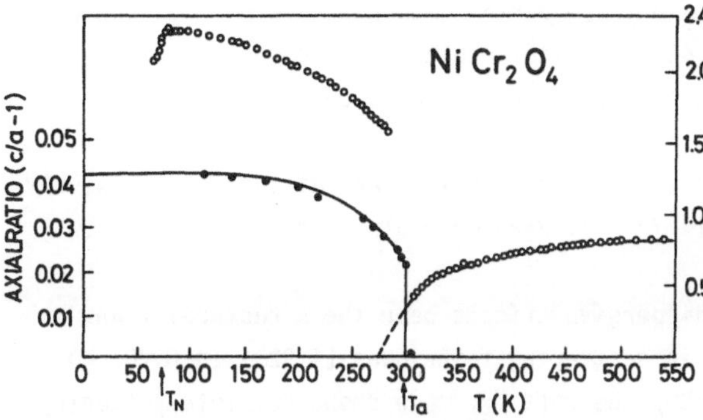

Fig.4.4. Sound velocity for $c_{11} - c_{12}$ mode (o) and c/a ratio (•) for $NiCr_2O_4$. Solid lines are theoretical fits to (4.29) with g' = 0 and (4.30) (from [4.86,87]

The temperature dependence of the soft mode and the order parameter can be quantitatively accounted for.

Another interesting transition metal compound studied in detail is $CsCuCl_3$ which undergoes a structural transition at T_a = 423 K. The $Cu^{2+}(^2D)$ ion in an octahedral environment has a Γ_3 ground state doublet. In the high-temperature $CsNiCl_3$ structure D^4_{6h}, the $CuCl_6$ octahedra are linked together by sharing faces, thus forming a one-dimensional chain structure along the hexagonal c axis. The low-temperature structure D^2_6 consists of helical displacements of the atoms from the high-temperature structure. This leads to an elongation of the octahedra with alternative axis. The atomic displacements are modulated by $\exp(i\underline{q}_0 \cdot \underline{r})$ with $\underline{q}_0 = (0,0,2\pi/3c)$ which is 1/3 of the reciprocal lattice vector.

This transition was investigated with optical, X-ray, thermal expansion, ultrasonic, and neutron scattering techniques [4.89,90]. No softening of a $\underline{q}_0$ phonon was observed, mainly because of the strong first-order nature of the transition. Elastic constant measurements gave the following results: The c_{33} mode (longitudinal wave propagating along the hexagonal c axis) exhibits a strong anomaly at the transition in the form of a step function, but it behaves normally in the high-temperature phase. The c_{44} mode (shear wave propagating along the c axis), however, shows strong softening in the high-temperature phase (measured from 530 to 423 K) which can be interpreted with (4.29). The fit gave again as in $NiCr_2O_4$, $g' \ll g^2/c_0$ and T_0 = 324 K. Therefore this magnetoelastic coupling to the bulk c_{44} strain competes with the Jahn-Teller coupling of an internal $\underline{q}_0$ mode and would lead to a structural transition at T_0. $CsCuCl_3$ is one of very few cases where the magnetoelastic coupling to macroscopic strains is not the dominant interaction.

Rare Earth Compounds: Rare earth compounds have transition temperatures of a few degrees Kelvin, whereas transition metal compounds have transition temperatures which are several hundred degrees. The differences are due to the same shielding effects which are responsible for the different strengths in crystal field parameters. All elastic experiments done on rare earth compounds have been discussed with the displaced phonon theory [4.43,84] which result in (4.28,29). We give a few typical experimental results and summarize the other compounds.

Rare Earth Vanadates: As already mentioned, these systems are discussed extensively in the review articles cited above [4.73,74]. Here we would like to point out two interesting facts.

The first is the observation of dispersive effects near the structural transition, using ultrasonic and Brillouin scattering techniques [4.92]. In $DyVO_4$ no such effects are observed, but in $TbVO_4$ the soft c_{66} mode shows complete softening for ultrasonic frequencies towards T_a = 34 K, whereas with increasing frequencies (6 - 16 GHz) the softening becomes less and less pronounced. This behavior is shown in Fig.4.5. It cannot be explained with the isothermal elastic constant expressions above, (4.6,29). In fact it implies that the order parameters can no longer follow the strain oscillations completely at the higher frequencies. One can describe the situation either by a phenomenological relaxation term, (4.8), or one has to solve the coupled equations of motion for the elastic wave and the order parameter (pseudospin). For $\omega\tau \gg 1$, with τ a local relaxation time, c_T will approach the background c_0 as explained in Sect.4.2.1. For the case of $TbVO_4$ different relaxation times (τ_{ss} spin-spin, τ_{sL} spin-lattice) and frequency and temperature regimes were introduced to describe the situation [4.92].

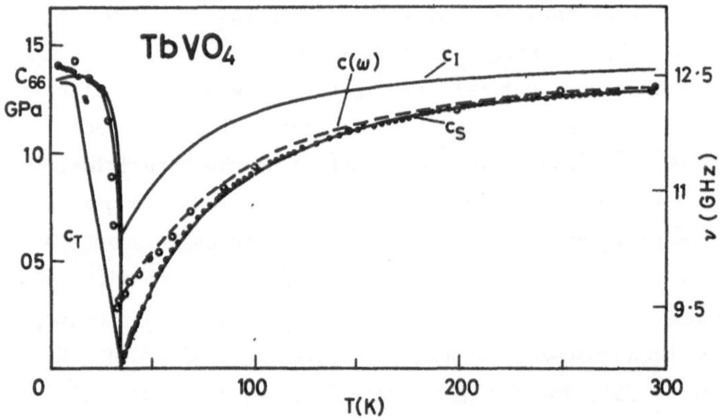

Fig.4.5. Ultrasonic (•) and Brillouin scattering (o) results for c_{66} mode for $TbVO_4$. The curves c_S, c_I, $c(\omega)$ are based on similar formulas as (4.29,30) (from [4.92])

The second point to discuss is the question of marginal or critical dimensionality. The rare earth vanadates, with a tetragonal to orthorhombic structural transition and a one-dimensional strain as order parameter, corresponding either to the c_{66} or to the $c_{11} - c_{12}$ mode, cannot have cubic invariants in the free energy. This is in contrast to the cubic case see (4.14) . The absence of cubic invariants makes a second-order phase transition possible and indeed the structural transitions in the vanadates seem to be of second order. In the vanadates (Dy, Tm, and $TbVO_4$) $d^* = 2$, [4.39]; therefore one expects mean-field behavior. This is what one finds for the temperature dependence of the soft mode [4.92,93], (4.29). One can test it also by measuring the order parameter using elastic modes which couple in higher order to the order parameter, (4.13) [4.94], or by optical birefringence experiments [4.95]. For $DyVO_4$ one finds a temperature dependence of the order parameter which is more Ising-like. This deviation from classical behavior indicates a strong optical phonon mode coupling [4.73].

Cubic Rare Earth Compounds: There are a great number of rare earth intermetallic compounds with NaCl or CsCl structures which exhibit structural phase transitions [4.80,96]. In the majority of cases the structural transition coincides with a magnetic transition. This has been tested in detail for some rare earth pnictides [4.78,80]. But there are a few cases where the structural and magnetic transition temperatures are different: the CsCl structure materials TmCd and TmZn. We briefly discuss these materials here [4.97,98], because these transitions were discovered and characterized mainly through elastic studies.

TmCd has a cubic-tetragonal phase transition at $T_a = 3.16$ K with no indication of a further magnetic transition down to 30 mK. On the other hand TmZn exhibits a structural transition at $T_a = 8.5$ K followed by a ferromagnetic one at 8.1 K. The $Tm^{3+}(4f^{12})$ $J = 6$ state has a Γ_5 ground state as verified by inelastic neutron scattering. In Fig.4.6 we show the soft modes $c_{11} - c_{12}$ for these two cases, together with a theoretical fit using (4.29). In these cases g' is no longer negligible, but is more than three times larger than g^2/c_0. This was also verified by magnetization and parastriction experiments. For CsCl structure materials there is no quadrupolar coupling to $q = 0$ optical phonons; therefore g' originates from conduction electron coupling [4.81]. These materials are good conductors and anomalies in the electric resistance due to the structural transition and crystal field levels have been observed. Therefore the aspherical Coulomb charge scattering is an effective mechanism for quadrupolar coupling. Another interesting property of the structural transition in TmCd and TmZn is the magnetic field dependence of T_a. With a magnetic field along the 100 direction one can shift T_a from 3.16 to 6.5 K at H = 30 kOe for TmCd and from 8.1 to 15 K at H = 20 kOe for TmZn. These effects arise via the Zeeman effect, which modifies the Tm^{3+} energy levels and wave functions, especially the Γ_5 ground state, and therefore also make the order parameter $\langle O_\Gamma \rangle$ magnetic field dependent. Recently the ultrasonic attenuation near the cooperative Jahn-Teller phase transition in TmZn was studied experimentally and theoretically [4.98].

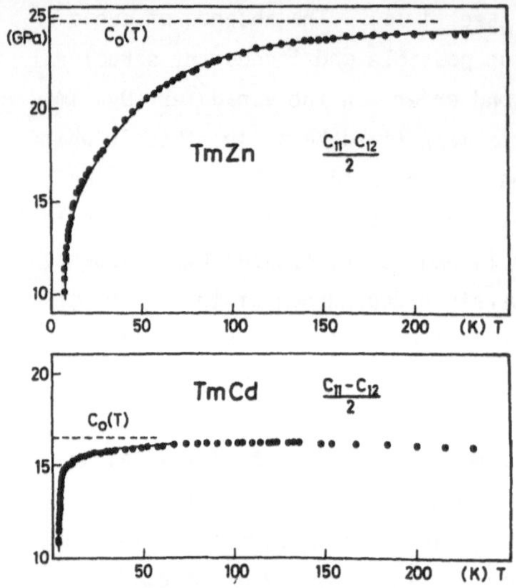

Fig.4.6. Shear stiffness $(c_{11}-c_{12})/2$ for TmCd and TmZn. Solid lines are fits using (4.29) [4.97,98]

Other Rare Earth Compounds: There are a number of other rare earth compounds which exhibit structural transitions. Notable among them are $PrAlO_3$ which is a well-characterized substance [4.99], but for which no ultrasonic studies have been performed. Another example is $PrCu_2$ which shows an induced cooperative Jahn-Teller transition at T = 7.5 K with strong softening of some symmetry modes (c_{55}) [4.100]. This transition has not yet been fully characterized.

Actinide Compounds: The only actinide compound which has been studied ultrasonically is UO_2. The c_{44} mode exhibits strong softening in the paramagnetic phase down to T_N = 65 K [4.101]. It has been interpreted in an analogous way as the transition metal compounds, using (4.29) or (4.6), respectively [4.83]. More recently internal displacements of the oxygen atoms have been found for $T < T_N$ [4.102] which, however, belong to $\underline{q} \neq 0$ optical phonon modes and therefore do not couple directly to the elastic modes.

d) *Band Jahn-Teller Transitions*

The β tungsten or *A 15 structure* compounds, such as Nb_3Sn, V_3Si, Nb_3Ge, Nb_3Ga, etc., are known as good superconductors. Some of them exhibit at low temperatures, but still in the normal state, a structural phase transition, cubic to tetragonal [4.103]. Many of the experimental and theoretical results have been reviewed before [4.104,105]. The temperature dependence of the Knight shift [4.106] and the magnetic susceptibility [4.107] have led to the idea that in these materials the Fermi level lies in a portion of the conduction band with a high and strongly varying density of states and that these features might also cause the phase transition. Such a peculiar band structure was explained by the crystallographic arrangement of transition metal atoms (Nb, V) in three sets of chains running along directions of the cube axis [4.108]. A one-dimensional tight-binding approximation gave a set of triply degenerate d electron bands with a singularity at the band edges. By the spontaneous strain developing in the tetragonal phase this degeneracy is lifted at the expense of lattice strain energy.

The shifts W_i of the band edges i (i = 1,2,3) from their original position in the high-temperature phase can be considered as components of the order parameter. Their coupling to the strain is linear and described by a deformation potential Ξ. Only strains that alter the atomic distances in different chains in a different way are considered, i.e., strain components e_{e1} and e_{e2}. Denoting by $N_i(E)$ the density of states in the ith subband, measured from the bandedge, and by n the total density of d electrons, the free-energy density of the d electron system can be written

$$F_{el} = nE_F - kT \sum_{i=1}^{3} \int_0^\infty N(E) \ln \left[1 + \exp\left(- \frac{E + W_i - E_F - \omega}{k_B T}\right)\right] dE \quad .$$

The effect of strain is taken into account by the individual band edge positions W_i and the variation ω of the Fermi level E_F as a consequence of the band shifts. By a strain e_{e2} (see Table 4.1) the bands are shifted as

$$W_1 = W_2 = - \Xi\, e_{e2}/2\sqrt{3} \quad \text{and} \quad W_3 = \Xi\, e_{e2}/\sqrt{3}$$

and the Fermi energy stays unchanged in lowest order of strain, $\omega = 0$.

In order to compare the formula for F_{el} with (4.3,4) we consider the effects of strain as sufficiently small and develop F_{el} around $W_i = 0$,

$$F_{el} = nE_F - 3k_B T \int_0^\infty N(E) \ln \left[1 + \exp\left(- \frac{E - E_F}{k_B T}\right)\right] dE$$
$$+ \frac{1}{2} \int_0^\infty N(E) \frac{\partial f(E,E_p)}{\partial E} dE \cdot \sum_{i=1}^{3} W_i^2 \quad .$$

The first two terms describe the d electron system without strain, i.e., F(Q) in (4.3); the third term corresponds to the coupling energy F_{int} with the condition for free motion of the order parameter, (4.5), already taken into account. The variation of the elastic stiffness $c_e = (c_{11} - c_{12})/2$ follows immediately from this term,

$$c_e(T) = c_e^0 + \frac{\partial^2 F_{el}}{\partial e_{e2}^2} = c_e^0 + \frac{1}{2} \Xi^2 \int_0^\infty N(E) \frac{\partial f(E,E_F)}{\partial E} dE \quad .$$

The integral, containing the derivative of the Fermi distribution function $f(E,E_F) = f(E - E_F)$, is negative, resulting in a softening effect of the electron system. Its magnitude depends largely on the electronic density of states at the Fermi level, but varies strongly with temperature if E_F is located in a region where N(E) shows large variations, i.e., close to a band edge. Both conditions, a large and strongly varying N(E) near the Fermi level, are met in some of the A 15 structure compounds.

The essential features of the elastic and magnetic properties, as well as the temperature dependence of the electric resistivity, can be discussed quantitatively in the simple and mathematically tractable model of a steplike density of states [4.109]

$$N(E) = N_0/3 \quad \text{for} \quad E > 0$$
$$= 0 \qquad\qquad E < 0 \quad .$$

The position of $E_F(T)$ as a function of temperature has to be calculated separately from the condition that the number of d electrons is conserved,

$$\int_0^\infty N(E)f(E,E_F)dE = \frac{1}{3} N_0 \, k_B T \ln\left[1 + \exp \frac{E_F}{k_B T}\right] = \frac{1}{3} N_0 E_F(0) \quad .$$

With the Fermi temperature $T_0 = E_F(T = 0)/k_B$ as a material parameter the elastic stiffness c_e follows.

$$c_e(T) = c_e^0 - \frac{1}{6} N_0 \, \Xi^2[1 - \exp(-T_0/T)] \quad . \tag{4.31}$$

The quantity in brackets rises continuously when T decreases to T = 0 and causes an elastic instability at $T_c^\sigma > 0$, if $N_0 \, \Xi^2/6 > c_e^0$.

Since the structural phase transition manifests itself so clearly in the elastic functions, this is another example where ultrasonic measurements can give direct access to the microscopic processes involved. The elastic stiffness functions were measured in single crystals of V_3Si [4.110] and Nb_3Sn [4.111,112]. In both materials a softening of $c_{11} - c_{12}$ was observed on approaching the transition temperature T_c = 21 K and 45 K, respectively, from above. In the tetragonal phase of Nb_3Sn the shear mode (110, 1$\bar{1}$0) could not be observed. From the variation of other sound velocities and under the assumption of no coupling to the dilatation, an increase of $c_{11} - c_{12}$ to the high-temperature value could be inferred. Also the other shear stiffness c_{44} decreases anomalously towards lower temperatures and changes its slope at T_c. This behavior cannot be explained by the theories discussed that neglect chain interactions. Figure 4.7 shows the measured temperature dependence of c_e and c_{44} in Nb_3Sn; the theoretical curve is based on the steplike density of states model with c_e^0 = 96.8 GPa, $N_0 \, \Xi^2/6$ = 121.3 GPa, and T_0 = 80 K. Inelastic neutron scattering measurements on Nb_3Sn [4.113] showed that the softening of the transverse acoustic branch (110,1$\bar{1}$0) extends to about a wave vector $q \approx 0.1\ a^*$ from the center of the Brillouin zone. In V_3Si the stiffness c_e stays at low values in the tetragonal phase.

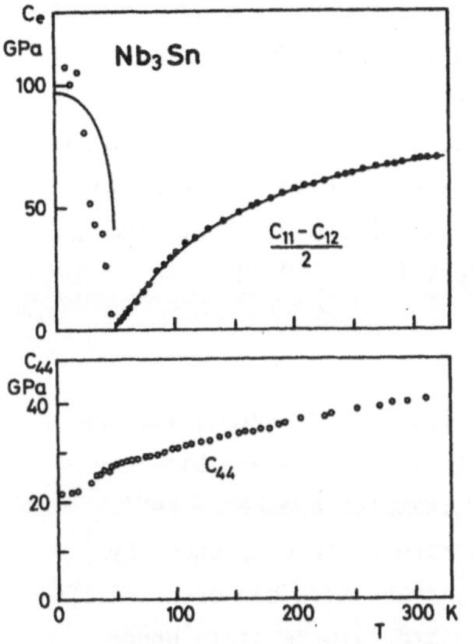

Fig.4.7. Shear stiffness $(c_{11} - c_{12})/2$ and c_{44} for Nb_3Sn. • Directly measured, o calculated from other modes assuming c_a = constant [4.112]

We now discuss some recent theoretical approaches, which bring in some new ideas. The anomalous temperature dependence of c_{44} in Nb_3Sn and band structure calculations [4.114] cast some doubt upon the idea of a one-dimensional electronic structure. A new model was proposed by GORKOV which places the Fermi level close to the X point of the Brillouin zone, where two single bands meet [4.115]. This degeneracy is removed when the transition metal atoms shift along the chains and form pairs. Analogous to the Peierls distortion (see Sect.4.3.2d) static charge-density waves develop and a gap opens up at the Fermi level. This atomic motion along the chains, belonging to an optical phonon mode E_g, is in turn linearly coupled to the corresponding strains e_{e1} and e_{e2}. Basically such a model still depends on a special, quasi-one-dimensional electronic band structure. Its basic ideas have been incorporated into a three-dimensional Landau model of charge-density waves [4.116]. In its most recent version Bhatt's theory [4.117] uses a tight-binding band structure with nearest and next-nearest neighbor interaction taken into account. In addition to the Peierls gap contribution, which is too small in a three-dimensional system to explain the elastic instability, a band Jahn-Teller type mechanism is acting. The necessary high density of states comes from saddle points at the M point and between the X and Γ points of the Brillouin zone. Recently another proposal was made to arrive at the necessary high and strongly varying electronic density of states [4.118]: the Fermi level located close to the R point, where the electron bands are sixfold degenerate.

In order to distinguish between the different models a direct experimental determination of the electronic density of states would be essential. Measurements of the variation of $c_{11} - c_{12}$ under a strong magnetic field can, in principle, give additional details in N(E), but suffer from technical difficulties and too low an accuracy [4.119-121].

Another interesting system, exhibiting a band Jahn-Teller coupling, is the CsCl structure $LaAg_xIn_{1-x}$ compounds [4.122]. Elastic constant measurements showed considerable precursor softening of the $c_{11} - c_{12}$ mode for LaAg and even stronger softening for the transforming compounds with x = 0.78, 0.89 [4.123]. A strain 5d band deformation potential coupling accounts quantitatively for the temperature dependence of the $c_{11} - c_{12}$ mode in the cubic phase for x = 1 and 0.89 but strong anharmonic coupling to soft zone boundary phonons occurs for $x \leq 0.78$.

e) *Mixed Valence Transitions*

Magnetic ions with unfilled 3d or 4f shells are often found in different valence states which compete for stability. The resulting state is said to exhibit interconfigurational fluctuations or valence fluctuations. Examples are $Fe^{2+} - Fe^{3+}$ ions in spinel compounds and $Sm^{2+} - Sm^{3+}$ ions in cubic rare earth compounds. By applying pressure or by alloying a valence instability transition can occur [4.124]. A famous example is SmS which collapses to an intermediate valence state under pressure or by alloying with YS, SmAs, etc. Large volume changes can be observed with no change in lattice symmetry, because the trivalent and divalent ions have quite different ionic radii. In a microscopic model describing the valence transition, one has to add the electronic and elastic energies and determine the order parameter and the soft mode in a similar way as we have outlined before. The appropriate macroscopic order parameter for such a transition is the volume strain $e_a = (V - V_c)/V_c$, the reduced volume, where V_c is the critical volume. The associated soft mode is then the bulk modulus c_a, which for cubic systems is (Table 4.1) $c_a = (c_{11} + 2c_{12})/3$. Three systems have been investigated so far elastically: $Sm_{1-x}Y_xS$, $Ce_{1-x}Th_x$, and TmSe.

In $Sm_{1-x}Y_xS$ room temperature elastic constant measurements [4.125] show a strong softening of c_a as a function of x for $x > x_c = 0.15$. With x increasing from 0 the system changes at x_c from a semiconducting phase with a $Sm(4f^6)$ configuration to a metallic phase with mixed valence $4f^6$ and $4f^5 5d^1$. For $x < x_c$, c_a is roughly constant, whereas for $x > x_c$, c_a diminishes for $x \to x_c$. For $x = x_c$ the temperature dependence of c_a has not yet been measured.

In $Ce_{1-x}Th_x$ and for x = 0.272 c_a softens as a function of T on approaching the valence instability transition temperature $T_0 = 148$ K. Actually in this experiment the Young modulus Y was measured and c_a estimated from it [4.126]. The resulting critical exponents for c_a or the compressibility turned out to be classical, as one would expect from the concept of marginal dimensionality (see Sect.4.2.4), $d^* = 0$ for the case of a valence instability.

In TmSe under atmospheric pressure, there is no valence instability transition. It is believed that TmSe consists of a mixture of Tm^{2+} and Tm^{3+}, exhibiting interconfigurational fluctuations. Consequently the elastic constants do not show any crystal electric field effects and c_a behaves normally [4.127]. However, as a function of stoichiometry (Tm_xSe_{1-x}) c_a exhibits again a minimum at $x = 0.5$ [4.128].

f) *Martensitic Phase Transformations*

The expression "martensitic transformation" applies in general to any diffusionless transformation in which the two phases are related by shear deformations. In this way, many of the examples given in this Sect.4.3.1 show also martensitic behavior. We would like to apply this term here in a narrower sense, by discussing only metal alloys whose constituents are not transition metal ions, e.g., no d band materials shall be considered here, only alloys with s and p bands.

In several cases it was observed that the $c_{11} - c_{12}$ mode shows considerable softening as a function of temperature or alloy composition. For example in indium thallium alloys, $In_{0.75}Tl_{0.25}$ and $In_{73}Tl_{27}$, the $c_{11} - c_{12}$ mode softens almost completely at 195 and 125 K, respectively [4.129]. The transformation in this case is from a face centered cubic to a face centered tetragonal phase. Another case are the $Au_xCu_{y-x}Zn_{1-y}$ alloys which transform from a body centered cubic structure to an orthorhombic one with $c_{11} - c_{12}$ softening for $x = 0.23$, $y = 0.53$ [4.130]. There are many more examples of such behavior (AgCd, CdMg alloys, etc.); however, the elastic results are less quantitative. Of interest is also a recent inelastic neutron scattering study of $AuCuZn_2$ which shows considerable softening of a transverse acoustic phonon, corresponding to the $c_{11} - c_{12}$ mode at small q [4.131]. The softening extends far into the Brillouin zone and it shows furthermore an anomalous structure at $q \sim 2/3 q_{BZ}(110) = \frac{a^*}{3}(110)$.

A theory for these martensitic phase transformations has been attempted recently, at least for the simple cubic cases as In-Tl and Cd-Mg alloys [4.132]. Starting point is a pseudopotential calculation which gives the interatomic potential $\Phi(R_{ij})$. The elastic constants can then be expressed, similarly as in the Fuchs theory of elastic constants, as combinations of first and second derivatives of this potential: $\tau = (1/R)(\partial\phi/\partial R)$, $R = \partial^2\phi/\partial R^2$. With the problem parametrized this way, one can in favorable cases see which term (τ or R) contributes most to the softening of a mode, say $c_{11} - c_{12}$. Since the theory has been formulated only for T = 0, one can discuss so far only softening as a function of alloy composition.

Concluding Remarks

In all the materials listed in this Sect.4.3.1 the strain is linearly related to some other microscopic quantity, also in cases where its nature is not fully known. This bilinear coupling produces a spontaneous strain in the ordered phase with a temperature variation characteristic for the order parameter (critical exponent β). Similarly the corresponding elastic functions $s_s = 1/c_s$ reflect the strong variations typical for the order parameter susceptibility (critical exponent γ). Therefore the strain and the quantities linearly related to it are both "primary order parameters" with equal rights. The notion "secondary" should be used only for quantities connected with higher powers of the primary order parameter.

Most of the transitions considered here are discontinuous (first order). The reason is that in these cases third-order invariants of the strain components involved can exist (4.14). Quoting critical exponents under these circumstances is rather dangerous. In the few examples, where a continuous transition might be present (TeO_2, Rochelle salt, $TbVO_4$, $PrAlO_3$), mean-field exponents with $\beta = 1/2$ and $\gamma = 1$ are encountered. This can be explained with the concept of marginal dimensionality discussed in Sect.4.2.4.

4.3.2 Strain Is Not Order Parameter

In the materials discussed here the strain couples to a power of the order parameter equal to or higher than two. Consequently the effects in the elastic stiffness functions near the phase transition are weaker than in the case of bilinear coupling, although a whole spectrum of fluctuations is involved. The order parameter may be a phonon coordinate or an orientational degree of freedom; in a ferroelectric it can be measured externally as a spontaneous electric polarization; in some low-dimensional metals it is a set of charge-density waves coupled with periodic lattice distortions; in the solid electrolytes (supraionic conductors) it describes the melting of one sublattice.

a) *Phonon Coordinate Is Order Parameter*

Here the critical fluctuations of the ordering quantity, being responsible for the effects upon the elastic stiffness and attenuation, are carried by the soft phonons. The range and the directional dependence of their correlation are closely connected with the dispersion in the critical branch in question. We discuss the cases of $SrTiO_3$, $Sn_xGe_{1-x}Te$, and SiO_2 in some detail.

Strontium Titanate ($SrTiO_3$)

This is the best known example in this class. It has a continuous, antiferrodistortive phase transition around 105 K. There is a coupling between all strain components and the square of the order parameter, which is the oxygen octahedra rotation angle, but the interaction with the dilatational strain h_a is rather weak [4.20,133].

$$F_{int} = h_a e_a (Q_1^2 + Q_2^2 + Q_3^2) + h_e\left[e_{e1}(Q_1^2 - Q_2^2) + e_{e2}(2Q_3^2 - Q_1^2 - Q_2^2)/\sqrt{3}\right] + h_t(e_4 Q_2 Q_3 + e_5 Q_3 Q_1 + e_6 Q_1 Q_2) \quad . \tag{4.32}$$

The temperature dependence of various sound velocities was measured by several authors [4.133-136]. Most of these measurements suffer from the fact that in the ordered, tetragonal phase domains appear whose motion influences the ultrasonic response. This explains the large scatter in the step size results and the coupling coefficients h_i derived from that. By applying biaxial compression, the formation of domains is suppressed, and more reliable results were obtained with a pronounced dip in the elastic functions close to T_c [4.136].

Stress experiments also helped to verify the idea, introduced by BRUCE and AHARONY [4.36] and discussed already in Sect.4.2.4, that in the stress-free state this transition represents a bicritical point, described by a three-dimensional Heisenberg model (n = 3), and that by applying uniaxial tension or compression the character of the phase transition can be modified to that of an Ising (n = 1) or x-y model (n = 2), respectively. The critical exponent for c_{11} was $\mu = 0.69$ in the unstressed state and decreased to 0.53 under biaxial compression of 117 bar and to 0.56 by uniaxial compression of 207 bar as shown in Fig.4.8a [4.137]. At the bicritical point the behavior of the elastic stiffness is determined by the axial crossover exponent

$$\phi_e = (2 + \mu - \alpha)/2 \sim 1.4 \quad . \tag{4.33}$$

Under stress μ should change gradually to different values valid for n = 2 or n = 1. Figure 4.8b shows a phase diagram in the pressure-temperature plane. The phase boundary varies linearly for hydrostatic strain (type A_g), but as a power law with the crossover exponent ϕ_e for symmetry-breaking strains of type E_g [4.36]. Since in the discussed pressure experiment both types of strain are generated, the indicated phase boundary is a superposition of the linear and the ϕ_e-dependent effect.

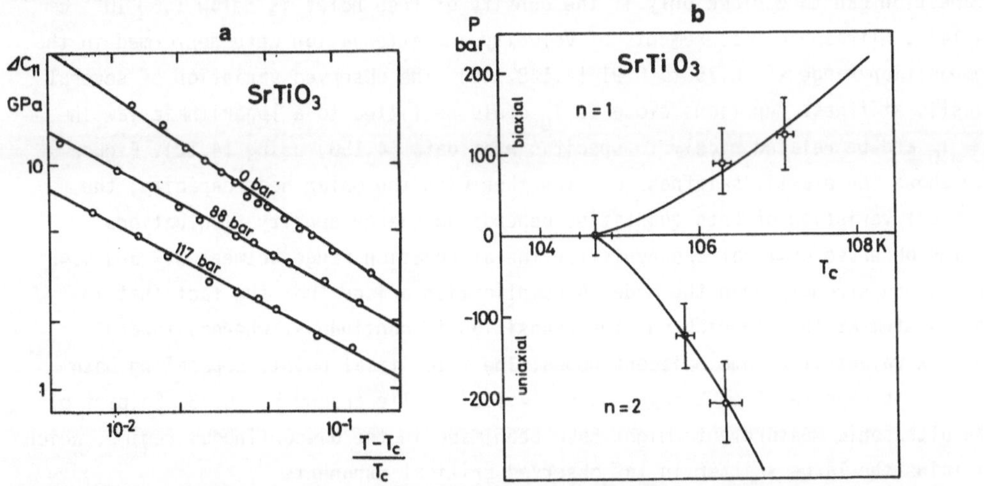

Fig.4.8. (a) Stiffness c_{11} of $SrTiO_3$ unstressed and under biaxial compression [4.137]. (b) Phase diagram of $SrTiO_3$ for biaxial and uniaxial compression. Data points from [4.137]. The solid line is a theoretical fit with $\phi_e = 1.4$ and $dT_c/dp = 1.8$ K/kbar for hydrostatic pressure

This example nicely shows that ultrasonic experiments are a valuable tool to locate the transition temperature and to determine phase diagrams. In a similar way the effect of hydrostatic pressure was measured [4.138], and a positive pressure coefficient $dT_c/dp = 1.8$ K/kbar was obtained in the low-pressure limit. Strontium titanate is an especially good candidate for the ultrasonic determination of T_c, a task that would otherwise need the introduction of paramagnetic impurities for electron-spin resonance or the large expense of a neutron scattering experiment. Recently the change in T_c due to oxygen vacancies and Nb doping was studied similarly [4.139].

The rise in the ultrasonic attenuation on approaching $T = T_c$ from above has been analyzed in terms of critical exponents. The results from various sources and from different modes scatter appreciably. For example, for the c_{11} mode the exponent varies between 0.65 and 2.2 [4.133,136,140,141]. The reason for this is probably that most of the measurements were made in polydomain samples. Besides this, any interpretation has to take into account that many of the sound modes investigated involve more than one critical exponent, as has been explained in Sects.4.2.3 and 4.2.4.

Tin-Germanium Telluride ($Sn_xGe_{1-x}Te$)

In the alloy system $Sn_xGe_{1-x}Te$ the transition temperature can be shifted from about 100 K for $x = 1$ to approximately 700 K for $x = 0$ by varying the composition x [4.144]. The ordering consists of a relative displacement of the anion (Te) and the cation (Sn,Ge) sublattices along the cubic [111] direction. It is carried by a soft optic phonon, which was observed in GeTe by Raman scattering [4.145] and in SnTe by inelastic neutron scattering [4.146]. In pure SnTe a complete phase transition can take place only if the density of free holes is below $1.5 \cdot 10^{20}$ cm^{-3} [4.147]. Ultrasonic measurements of velocity and attenuation were performed in the composition range $x = 0.75$ to 0.91 [4.148,149]. The observed variation of several elastic stiffness functions close to T_0 could be fitted to a logarithmic law in $T - T_0$ and be related nicely to specific heat data [4.150] using (4.12). Figure 4.9 shows the elastic stiffness c_{11} together with the molar heat capacity; the critical variation of both quantities depends on energy density fluctuations.

The observed critical exponents for the attenuation range between 0.7 and 4.4 and depend strongly upon the mode. A complication arises from the fact that in this system at the tin-rich end the transition is continuous, whereas towards lower x values it becomes discontinuous. The tricritical point, separating both ranges, is expected from X-ray studies [4.151] to lie around $x = 0.73$. So part of the ultrasonic measurements might have been made in the discontinuous regime, which explains the large scatter in the observed critical exponents.

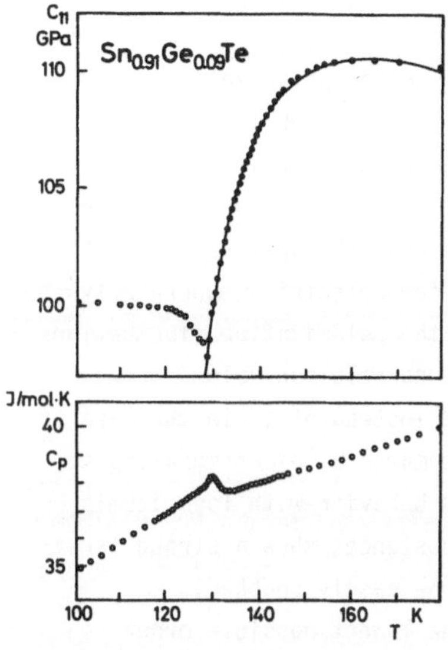

Fig.4.9. Stiffness c_{11} of $Sn_{0.91}$ $Ge_{0.09}$ Te compared with the molar heat capacity near the phase transition [4.148,150]

Quartz (SiO_2)

In quartz, the temperature variation of the elastic functions around the hexagonal-trigonal transition at 846 K (α - β transition) has been known since 1941 [4.152, 153]. Recently there has been an interest in deducing the variation of the order parameter from elastic functions [4.154,155]. The function c_{14} is zero in the hexagonal phase for symmetry reasons and increases proportional to the order parameter in the trigonal phase. Such a behavior is described by an interaction energy of the form

$$F_{int} = \kappa Q(e_5e_6 - e_1e_4 + e_2e_4) \quad . \qquad (4.34)$$

This relationship follows from the decomposition of the direct product of the representations E_1 for (e_5, - e_1) and E_2 for (e_6, e_1 - e_2) of the point group D_6. It contains the representation B_1 of the order parameter, a phonon coordinate that roughly describes the rotation of the SiO_4 tetrahedra. On the other hand, c_{66} couples with the square of the order parameter according to $F_{int} = \lambda Q^2e_6^2$ (4.13). An evaluation was tried and comparison was made with values for $\langle Q \rangle$ from other sources. As a consequence of the discontinuous nature of the phase transition a fit to critical laws is ambiguous in this case. Indeed, critical exponents β = 1/3 and 1/6 were found, depending on the number of fitting parameters and the temperature range considered [4.155].

Other systems, belonging to this class of materials, such as *potassium manganese fluoride* ($KMnF_3$), *gadolinium molybdate* [$Gd_2(MoO_4)_3$], *terbium molybdate* [$Tb_2(MoO_4)_3$], and *barium manganese fluoride* ($BaMnF_4$) we list with appropriate references in Table 4.4.

b) *Electric Polarization Is Order Parameter*

In this section materials are discussed in which the electric polarization is a primary order parameter, but the strain is coupled in higher order to it. Depending on $q_0 = 0$ or $\neq 0$ these crystals appear as ferroelectrics or antiferroelectrics in the ordered phase. The fact that the ordering quantity carries along the long-range electric dipole field has important consequences upon the spatial correlations in the polarization fluctuation spectrum. In uniaxial ferroelectrics, where only one polarization component exists, the correlation of the polarization fluctuations is strongly enhanced along the ferroelectric axis. The correlation volume grows with ξ^4 (ξ = correlation length) when T_0 is approached instead of ξ^3 in the case of isotropic short-range interactions. This decreases the marginal dimensionality to $d^* = 3$, i.e., in a three-dimensional system mean-field behavior with logarithmic corrections is to be expected. However, most of the substances show a strong first-order transition and therefore these arguments cannot be easily checked.

With respect to the polarization-strain coupling the lowest possible order is linear in the strain and quadratic in the polarization (electrostrictive coupling). Bilinear coupling was discussed in Sect.4.3.1. The general behavior of the elastic properties is a decrease of certain stiffness functions in form of a step or a dip at T_c. The corresponding attenuation increases on both sides of the transition point T_c. As a typical example to be discussed in detail we choose triglycine sulfate (TGS) and barium titanate.

Triglycine Sulfate and Selenate $[(NH_2CH_2COOH)_3H_2SO_4, (-)_3H_2SeO_4]$

These ferroelectrics, usually abbreviated as TGS and TGSe, exhibit phase transitions at 322.6 K and 295 K, respectively, that are considered as second order. Therefore a great number of investigations have been concentrated on them with the goal in mind of verifying experimentally the logarithmic corrections in certain critical quantities as predicted by theory. Ultrasonic measurements, both of sound velocities and attenuation, have contributed significantly to this question.

Following theoretical treatments [4.161-163] a nearly logarithmic temperature dependence of the specific heat, e.g., above T_c, is expected

$$\Delta C(T) = \frac{1}{3}\Delta C_0\left\{[1 + 3b\ \ln(\varepsilon_0/\varepsilon)]^{1/3} - 1\right\}$$

$$\approx \frac{1}{3}\Delta C_0 b\ \ln(\varepsilon_0/\varepsilon)$$

with $\varepsilon = (T - T_c)/T_c$, ε_0 and b as nonuniversal material parameters, and ΔC_0 as the jump in C at T_c. A similar formula holds for $T < T_c$. According to (4.18) the elastic stiffness varies like $\Delta C/T$ and should show practically the same logarithmic dependence. Recent accurate velocity measurements on TGS by STRUKOV et al. [4.164, 165] and by TODO [4.166] on TGSe could verify this. TODO was able to fit his data for c_{11}, c_{22}, and c_{33} to a logarithmic law in ε over nearly three decades. The

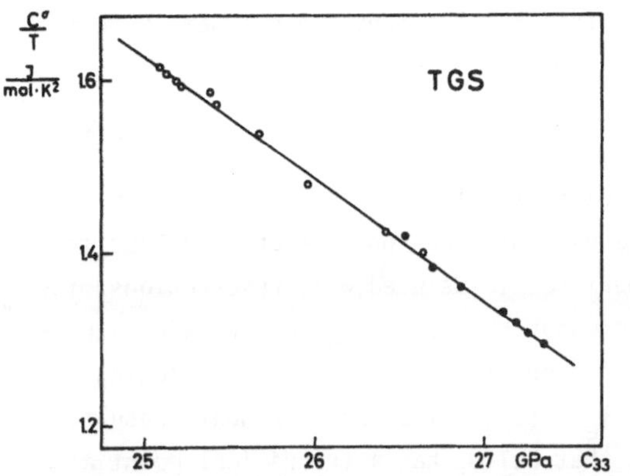

Fig.4.10. Total molar heat capacity divided by T as a function of elastic stiffness c_{33} for TGS [4.165]

relation between elastic functions and specific heat was well demonstrated [4.165]: Figure 4.10 (from [4.165]) gives C(T)/T as a function of c_{33}; the linear relationship verifies (4.18) and the theoretical Ansatz treating the effect as coupling between strain and energy-density fluctuations.

The ultrasonic attenuation in both substances was found to vary like $(\ln\varepsilon)^2$ [4.162,166,167]. From Sect.4.2.3 it follows that the responsible relaxation time varies also like the specific heat. Such a result was explained first by KAWASAKI [4.168]. The characteristic time τ_e for energy-density fluctuations involves, besides the volume specific heat C, the thermal conductivity λ and a kinetic coefficient Λ_e, analogous to (4.10)

$$\tau_e(q) = C/(\lambda q^2 + \Lambda_e) \quad .$$

For ultrasonic frequencies the λ contribution is negligible. Similar variations for the longitudinal wave attenuation were found in sodium nitrite [4.169,170], dicalcium strontium propionate [4.171], dicalcium lead propionate [4.172], and antimony sulfide iodide [4.173], all uniaxial ferroelectrics.

The attenuation for longitudinal waves in TGSe in the ordered phase has a pronounced anisotropy in the bc plane [4.174]. By the dipole-dipole interaction spatial polarization fluctuations are strongly suppressed when the wave vector points into the polarization direction (b direction). In the ordered phase the attenuation has a contribution from quasilinear coupling to the polarization fluctuations (Landau-Khalatnikov type), which becomes smaller when the wave vector gets a component parallel to the b direction. By varying the propagation direction it is, in principle, possible to separate the relaxation part (Landau-Khalatnikov) from the quadratically coupled fluctuation contribution. A quantitative evaluation could check the validity of a universal relation between both contributions [4.162, 163].

In *barium titinate* ($BaTiO_3$), the best-known example of a displacive ferroelectric, the variation of c_{11} and c_{44} was measured in the vicinity of the upper transition (cubic to tetragonal) at $T_c = 401$ K [4.175]. The function c_{11} was found to vary above T_c as

$$c_{11}(T) = c_{11}^0 - A_1(T - T_0)^{-\mu} \tag{4.35}$$

with the paraelectric Curie temperature (lower stability limit) $T_0 = 384$ K and the critical exponent $\mu = 0.41$. Although for a discontinuous phase transition, as is the case here, the notion of a critical exponent has to be used with reservations, its magnitude suggests a connection with the crossover exponent ϕ_e for uniaxial stress in a similar way as for $SrTiO_3$ at its antiferrodistortive transition. Assuming mean-field behavior, a crossover exponent $\phi_e = (\mu - \alpha + 2)/2 = 1.2$ would result. Another complication arises from the fact that $BaTiO_3$ has a tricritical point at a hydrostatic pressure of (34 ± 2) kbar and 291 K [4.176]. In order to check its effect upon the elastic behavior the bulk modulus $(c_{11} + 2c_{12})/3$ would have to be determined.

Other Systems

A group of materials, recently studied in Japan, are the *dicalcium propionates* of Sr, Pb, and Ba [4.171,172]. They display various phases, starting from cubic O and going to tetragonal C_4. Some of the phases are ferroelectric. Two features are worth mentioning: in $Ca_2Sr(C_2H_5COO)_6$ and $Ca_2Pb(C_2H_5COO)_6$ the longitudinal wave velocity in the c direction could well be described by a logarithmic law in $T - T_c$, characteristic for uniaxial ferroelectrics. Secondly, in $Ca_2Sr(C_2H_5COO)_6$ a noticeable dispersion in the longitudinal sound velocity in the c direction was observed for frequencies between 3 and 30 MHz, indicating a rather slow process with which the ultrasonic strain interacts [4.177].

Lead Germanate ($Pb_5Ge_3O_{11}$) has recently been studied by ultrasonic [4.178,179], light [4.180], and neutron scattering [4.182]. In the elastic stiffness functions c_{11}, c_{12}, and c_{33} measured ultrasonically a dip has been observed, whereas c_{44} and $c_{66} = (c_{11} - c_{12})/2$ do not show any variations at T_c [4.178,179]. The transition at $T_c = 450$ K is connected with the loss of the horizontal mirror plane only (point groups $C_{3h} - C_3$). The order parameter P_3 is of symmetry type A"; its square can couple only to strains of type A', i.e., to $e_1 + e_2$ and e_3. The combinations of stiffness functions affected by the phase transition are $c_{11} + c_{12}$, c_{33}, and c_{13}. The attenuation observed in connection with $c_{11} + c_{12}$ and c_{33} was analyzed in a way as it is done for piezodistortive materials.

Other systems such as sodium nitrite ($NaNO_2$) and antimony sulfur iodide (SbSI) are listed in Table 4.5 with appropriate references. The incommensurate phase of $NaNO_2$ is the subject of a recent study [4.170].

c) *Orientational Degree of Freedom Is Order Parameter*

Ammonium halides (NH_4Cl), (NH_4Br), and (NH_4I) undergo order-disorder phase transitions [4.181]. For example, NH_4Cl is a solid with CsCl structure below 456 K. With the N atom at the body center, the N-H bonds point towards the Cl ions along [111] directions. Therefore there are two possible orientations of these NH_4 tetrahedra in the high-temperature phase, which are occupied with equal probabilities. Below T_c = 242 K the tetrahedra become ordered. An interaction between ammonium groups is necessary for that. At least part of this interaction is electrostatic in nature, and the lowest nonvanishing term in a multipole expansion, which depends on the relative orientation of the tetrahedra, is the octupole-octupole interaction, leading to a parallel ordering of neighboring NH_4 groups. A second type of interaction is indirect and goes via polarizability of the halide ion. This octupole-dipole-octupole coupling leads to an antiferrodistortive ordering. These two forces are nearly balanced, and the type of anion determines which of both dominates: NH_4Cl orders ferrodistortively, NH_4Br antiferrodistortively [4.183,184]. Because of the two possible orientations of the ammonium ions one can associate pseudospin variables $\sigma = \pm 1$ to the positions. In a simple model the interaction is of the Ising type

$$H_I = -\frac{1}{2}\sum_{i,j} J_{ij}\sigma_i\sigma_j \quad . \tag{4.36}$$

Because there is no crystal symmetry change at T_c for NH_4Cl, except the loss of the center of inversion, the coupling to strain can be described by the volume dependence of $J_{ij} = J(V) = J(e_a)$.

At atmospheric pressure the phase transition in NH_4Cl is slightly first order; under hydrostatic pressure it changes its character and becomes second order as observed with ultrasonic velocity and length measurements [4.185]. At p = 1400 bar and T = 254 K, a multicritical, possibly tricritical, point exists. The critical exponents of the various thermodynamic quantities, such as specific heat, thermal expansion, and the temperature dependence of the order parameter, determined by neutron Bragg reflection, do not clearly discriminate between either a tricritical or fourth-order Gaussian point [4.186]. Therefore many basic aspects of this transition still remain unsettled and NH_4Cl is not necessarily a representative example for a compressible Ising model [4.187].

Ultrasonic velocity and attenuation measurements have been performed in NH_4Cl as a function of pressure and temperature in the MHz frequency range [4.181]. In addition some velocities were determined at atmospheric pressure by Brillouin scattering at about 20 GHz. Only longitudinal waves exhibit velocity dispersion $v(\omega) - v(0)$ and attenuation anomalies near the phase transition. This confirms the statement made at the beginning, that the interaction between strain and ordering quantity is a consequence of the volume dependence of the interaction parameter J, a situation analogous to the elastic behavior at magnetic phase transitions.

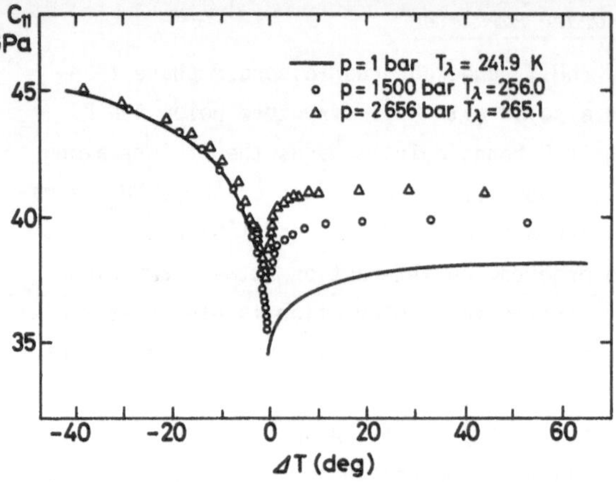

Fig.4.11. Elastic stiffness c_{11} as a function of pressure and temperature difference $\Delta T = T - T_c(p)$ for NH_4Cl [4.181]

In Figure 4.11 we show data for the elastic stiffness c_{11} as a function of pressure and $\Delta T = T - T_c(p)$. One notices pressure-dependent changes in the high-temperature phase, but almost no pressure dependence in the ordered phase. The ultrasonic attenuation is described by critical exponents $\rho = 1.1$ at high pressures and $\rho = 0.9$ at 1 bar. Using (4.20) from mode-mode coupling theory and the observed specific heat exponent $\alpha = 0.5$ [4.186], an anomalously low dynamical critical exponent $z = 0.8$ to 1.2 results.

d) *Charge-Density Wave Transitions*

In low-dimensional structures the coupled electron-lattice system can become unstable against a periodic distortion of the lattice, accompanied by a periodic modulation of the electronic charge density and the formation of a gap at the Fermi level. In three-dimensional metals so-called Kohn anomalies are known to occur in the phonon spectrum for wave vectors equal to twice the Fermi wave vector k_F. In two- and one-dimensional systems much larger sections of the Fermi surface contribute and can lead to a complete instability against such a Peierls distortion. The resulting static charge-density waves can be commensurate or incommensurate to the lattice periodicity, giving rise to a large variety of phenomena.

Ultrasonic studies of this type of phase transitions have been rather scarce in the past. In the one-dimensional charge-transfer salt *tetrathiafulvalene tetracyanoquinodimethane* (TTF-TCNQ) a change in slope at about 50 K was observed in the temperature variation of several sound velocities [4.189]. Between 50 and 0 K the velocities increase by about 1.5%, compared to their extrapolated values, as obtained from the variation above 50 K. This effect is explained as a softening of the elastic stiffness by the electron transfer between the TTF and TCNQ electronic bands under the action of strain, similar to the processes in A 15 compounds. Below 50 K a gap forms at the Fermi level, and the electron transfer is suppressed.

The mixed-valence compound *potassium tetracyanoplatinate* [$K_2Pt(CN)_4Br_{0.3}.3.2H_2O$ (KCP)] behaves in a similar way like a quasi-one-dimensional conductor. Ultrasonic measurements of sound waves along the a and c axis [4.190] indicate a considerable softening of c_{66} and c_{44} in the temperature range between 30 and 80 K. Its origin, as well as that of the attenuation, peaking in this temperature range, are not yet understood. The resonance of flexural vibrations gave values for Young's modulus $1/s_{33}$ showing a weak dip at about 40 K [4.191].

In the layered *transition-metal dichalcogenides* ($TaSe_2$) and ($NbSe_2$) the elastic compliance s_{11} was measured by the resonance of flexural vibrations [4.192]. Anomalies appear at the onset of incommensurate charge-density waves (at 122 K in 2H-$TaSe_2$ and at 29.8 K in 2H-$NbSe_2$) and at the discontinuous transition to the commensurate state at 90 K in 2H-$TaSe_2$. Ultrasonic pulse echo experiments with transverse sound waves propagating perpendicular to the layers of 2H-$NbSe_2$ show a similar temperature variation in the velocity [4.193]. The study of this group of materials has just begun, and more results and progress can be expected, especially on the microscopic mechanism of the strain order parameter coupling. We mention a recent study on perovskite-type layer structures, $(CH_3NH_3)_2\ MnCl_4$(=MAMC) [4.194].

e) *Solid Electrolytes*

Solid electrolytes, also called superionic conductors, recently received great technical interest. There exist excellent reviews on this topic; see, e.g. [4.195, 196]. These crystals are distinguished by a large ionic conductivity within one sublattice and sometimes exhibit also structural phase transitions. In *rubidium tetrasilver pentaiodide* ($RbAg_4I_5$) the elastic functions c_a, c_e, and $c_t = c_{44}$ and the corresponding attenuations were determined in the cubic high-temperature phase and around the phase transition at 208 K [4.197,198]. The variation of c_a indicates a coupling between strain and order parameter squared, although the nature of the order parameter is not known. The stiffness c_{44} behaves similarly with an additional steep downward dip at T_c. Both modes are connected with a critical rise of the attenuation on both sides of T_c. The shear stiffness $c_e = (c_{11} - c_{12})/2$ has in its temperature variation only a change in slope at T_c; no attenuation is observed. A second discontinuous phase transition at 122 K announces itself by a strong rise in both shear wave attenuations, but could not be investigated ultrasonically, because the crystals cracked. Ultrasonic studies of solid electrolytes should in principle give important information on relaxation processes connected with the sublattice ion mobilities. A theoretical interpretation of the elastic properties of $RbAg_4I_5$ based on a pseudospin model was given recently [4.208].

f) *Incommensurate Phases*

After completion of this article, incommensurate phases have become a subject of considerable interest. These phases are characterized by a wave vector q_0 that is not a rational fraction of a reciprocal lattice vector. Frequently there is a phase transition at lower temperatures to a phase in which q_0 is locked in at a nearby commensurate value. The new excitations in the incommensurate phase are amplitude and phase fluctuations of the ordering coordinate. The coupling of the strain to these excitations (ampliton and phason) is always of higher order.

Some materials undergoing incommensurate phase transitions have already been mentioned before: $NaNO_2$, $BaMnF_4$, and transition-metal dichalcogenides. New cases, where the elastic properties have been studied, are potassium selenate (K_2SeO_4) and its isomorphs and dipotassium lead hexanitrocuprate [$K_2PbCu(NO_2)_6$]. These new substances have been added in Table 4.6.

4.4 Summary

Our tables of materials studied by ultrasonic techniques around their structural phase transition points contain more than 40 entries. This list is perhaps not complete and will certainly have to be amended by the time this volume is published. Including other types of phase transitions, such as magnetic ordering, the lambda point in liquid helium, gas-liquid transitions, and the various transition phenomena occurring in liquid crystals, would multiply this number of materials. It definitely shows that the measurement of elastic properties in the vicinity of phase transformations is a valuable tool to probe more or less directly the processes driving the phase transition. Similar to practically all other experimental methods used for this purpose they should not stand alone, but be used in conjunction with other complementary experiments. Only in this way can they give their maximum yield.

The general results derived from this survey can be summarized as follows:

1) All the processes that couple to the strain show up in some way in the elastic functions, i.e., the temperature variation of the sound velocities.

2) In cases where this coupling is bilinear, the effect is always strong and drives the system into an elastic instability at the stability limit of the phase transitions. The strong variation of the order parameter susceptibility appears directly in the elastic response. The acoustic strain field directly probes the ordering quantity at the frequency and wave number of the sound wave.

3) Different mechanisms develop similar effects in the elastic functions if the type of coupling is formally the same.

4) The ultrasonic attenuation can give information on the dynamic response of the ordering quantity. In the case of bilinear coupling this happens for the fre-

quency and wave number of the probing sound wave; for higher order coupling this refers to an average over the whole relevant fluctuation spectrum. Experimentally, however, it is far from clear that the attenuation anomalies are due to critical fluctuations. Similarly as in the central peak problem for neutron-scattering, inhomogeneities, impurities, and domains can contribute to a rise in attenuation which may have very little to do with the physical processes described in (4.16,18, 19). Therefore we emphasized the temperature dependence of the various elastic functions in this review and only cautiously refered to the attenuation results.

References

4.1 H.J. McSkimmin: J. Acoust. Soc. Am. *33*, 12 (1961)
H.J. McSkimmin, P, Andreatch: J. Acoust. Soc. Am. *34*, 609 (1962)
4.2 T.J. Moran, B. Lüthi: Phys. Rev. *187*, 710 (1969)
4.3 R.D. Holbrook: J. Acoust. Soc. Am. *20*, 590 (1948)
N.P. Cedrone, D.R. Curram: J. Acoust. Soc. Am. *26*, 963 (1954)
4.4 E.P. Papadakis: J. Appl. Phys. *35*, 1474 (1964)
4.5 E.P. Papadakis: In *Physical Acoustics*, Vol.12, ed. by W.P. Mason, R.N. Thurston (Academic, New York 1976) p.277
4.6 R. Truell, C. Elbaum, B.B. Chick: *Ultrasonic Methods in Solid State Physics* (Academic, New York 1969)
4.7 E.R. Fuller, A.V. Granato, J. Holder, E.R. Naimon: In *Methods of Experimental Physics*, Vol.11, ed. by R.V. Coleman (Academic, New York 1974) p.371
4.8 D.I. Bolef: In *Physical Acoustics*, Vol.IV, part A, ed. by W.P. Mason, R.N. Thurston (Academic, New York 1966) p.113
4.9 D.I. Bolef, J.G. Miller: In *Physical Acoustics*, Vol.VIII (Academic, New York 1971) p.95
4.10 T.A. Read, C.A. Wert, M. Metzger: In *Methods of Experimental Physics*, Vol.6A, ed. by K. Lark-Horovitz, V.A. Johnson (Academic, New York 1974) p.291
4.11 W.P. Mason: *Piezoelectric Crystals and Their Application to Ultrasonics* (Van Nostrand, Princeton 1950)
4.12 G. Rupprecht, W.H. Winter: Phys. Rev. *155*, 1019 (1967)
4.13 J. de Klerk: In *Physical Acoustics*, Vol.IV, part A, ed. by W.P. Mason (Academic, New York 1966) p.195
4.14 J.D. Llewellyn, H.M. Montagu-Pollock, E.R. Dobbs: J. Phys. E*2*, 535 (1969)
4.15 Ch. Frenois, J. Joffrin, A. Levelut: J. Phys. (Paris) *37*, 275 (1976)
R.M. Holt, K. Fossheim: Ferroelectrics *25*, 515 (1980)
4.16 P.J. Debye, F.W. Sears: Proc. Nat. Acad. Sci. *18*, 409 (1932)
4.17 L. Bergmann, C. Schäfer: Z. Tech. Phys. *17*, 441 (1936)
4.18 A.F. Levites, K.A. Minaeva, B.A. Strukov, V.I. Teleshevskii: Prib. Tekh. Eksp. *8*, 187 (1974) [English transl.: Instrum. Exp. Tech. (USSR) *17*, 1473 (1974)]
4.19 L.D. Landau, I.M. Khalatnikov: Dokl. Akad. Nauk. USSR *96*, 469 (1954)
4.20 J.C. Slonczewski, H. Thomas: Phys. Rev. B*1*, 3599 (1970)
4.21 E. Pytte: Phys. Rev. B*1*, 924 (1970)
4.22 A. Michelson: Phys. Rev. B*14*, 4121 (1976)
4.23 N. Boccara: Ann. Phys. (N.Y.) *47*, 40 (1968)
4.24 R.N. Thurston, K. Brugger: Phys. Rev. *133*A, 1604 (1964)
4.25 D. Mukamel, S. Krinsky, P. Bak: AIP Conf. Proc. *29*, 474 (1975)
4.26 D. Mukamel, S. Krinsky: J. Phys. C*8*, L 496 (1975)
4.27 K. Kawasaki: "Mode Coupling and Critical Dynamics", in *Phase Transitions and Critical Phenomena*, Vol.5a, ed. by C. Domb, M.S. Green (Academic, London 1976) pp.165-403
4.28 L.D. Landau, E.M. Lifshitz: *Statistical Physics* (Pergamon, London 1959) pp.398 ff.

4.29 K.K. Murata: Phys. Rev. B*13*, 4015 (1976)
4.30 F. Schwabl: Phys. Rev. B*7*, 2038 (1973)
4.31 K. Kawasaki: Int. J. Magn. *1*, 171 (1971)
4.32 L.P. Kadanoff, J. Swift: Phys. Rev. *166*, 89 (1968)
4.33 B.I. Halperin, P.C. Hohenberg: Phys. Rev. *177*, 952 (1969)
4.34 M.E. Fisher: Rev. Mod. Phys. *46*, 597 (1974)
4.35 P.C. Hohenberg, B.I. Halperin: Rev. Mod. Phys. *49*, 435 (1977)
4.36 A.D. Bruce, A. Aharony: Phys. Rev. B*11*, 478 (1975)
4.37 J. Als-Nielsen, R.J. Birgenau: Am. J. Phys. *45*, 554 (1977)
4.38 A. Aharony, B.I. Halperin: Phys. Rev. Lett. *35*, 1308 (1975)
4.39 R.A. Cowley: Phys. Rev. B*13*, 4877 (1976)
4.40 R. Folk, H. Iro, F. Schwabl: Z. Phys. B*25*, 69 (1976)
4.41 K. Aizu: Phys. Rev. B*2*, 754 (1970)
4.42 K.K. Kobayashi: J. Phys. Soc. Jpn. *24*, 497 (1968)
4.43 R.J. Elliott, A.P. Young, S.R.P. Smith: J. Phys. C*4*, L 317 (1971)
4.44 W.P. Mason: Phys. Rev. *69*, 173 (1946)
4.45 C.W. Garland, D.B. Novotny: Phys. Rev. *177*, 971 (1969)
4.46 E. Litov, C.W. Garland: Phys. Rev. B*2*, 4597 (1970)
4.47 E. Litov, E.A. Uehling: Phys. Rev. B*1*, 3713 (1970); Phys. Rev. Lett. *21*, 809 (1968)
4.48 E.M. Brody, H.Z. Cummins: Phys. Rev. B*9*, 179 (1974); Phys. Rev. Lett. *21*, 1263 (1968)
4.49 H.B. Silsbee, E.A. Uehling, V.H. Schmidt: Phys. Rev. *133*, A 165 (1964)
4.50 Y. Tuval, R.E. Nettleton: J. Phys. C*9*, 1159 (1976)
4.51 E. Harnik, M. Shimshoni: Phys. Lett. *29*A, 620 (1969)
4.52 G.P. Singh, B.K. Basu: Phys. Lett. *64*A, 425 (1978)
4.53 R.E. Nettleton, Y. Tuval: J. Phys. C*10*, 3179 (1977)
4.54 R.J. Pollina, C.W. Garland: Phys. Rev. B*12*, 362 (1975)
4.55 R. Kawashima, I. Tatsusaki: J. Phys. Soc. Jpn. *42*, 564 (1977)
4.56 W.J. Price: Phys. Rev. *75*, 946 (1949)
4.57 H. Suga, T. Matsuo, S. Seki: Bull. Chem. Soc. Jpn. *38*, 1115 (1965)
4.58 K.H. Michel, J. Naudts: Phys. Rev. Lett. *39*, 212 (1977)
4.59 K.H. Michel, J. Naudts: J. Chem. Phys. *67*, 547 (1977)
4.60 S. Haussühl: Solid State Commun. *13*, 147 (1973)
4.61 W. Rehwald, J.R. Sandercock, M. Rossinelli: Phys. Status Solidi A*42*, 699 (1977)
4.62 W. Krasser, U. Buchenau, S. Haussühl: Solid State Commun. *18*, 287 (1976)
4.63 M. Boissier, R. Vacher, D. Fontaine, R.M. Pick: J. Phys. (Paris) *39*, 205 (1978)
4.64 P.S. Peercy, I.J. Fritz, G.A. Samara: J. Phys. Chem. Solids *36*, 1105 (1975)
4.65 P.S. Peercy, I.J. Fritz: Phys. Rev. Lett. *32*, 466 (1974)
4.66 D.B. McWhan, R.J. Birgeneau, W.A. Bonner, H. Taub, J.D. Axe: J. Phys. C*8*, L 81 (1975)
4.67 E.F. Skelton, J.L. Feldman, C.Y. Liu, I.L. Spain: Phys. Rev. B*13*, 2605 (1976)
4.68 T.G. Worlton, R.A. Beyerlein: Phys. Rev. B*12*, 1899 (1975)
4.69 R. Englman: *The Jahn-Teller-Effect in Molecules and Crystals* (Wiley Interscience, London 1972)
4.70 F. Ham: In *Electronic Paramagnetic Resonance*, ed. by S. Geschwind (Plenum, New York 1972)
4.71 J.D. Dunitz, L.E. Orgel: J. Phys. Chem. Solids *3*, 20 (1957)
4.72 D.S. McClure: J. Phys. Chem. Solids *3*, 311 (1957)
4.73 G.A. Gehring, K.A. Gehring: Rep. Prog. Phys. *38*, 1 (1975)
4.74 R.L. Melcher: In *Physical Acoustics*, Vol.XII, ed. by W.P. Mason, R.N. Thurston (Academic, New York 1976) p.1
4.75 H. Thomas: In *Electron-Phonon Interactions and Phase Transitions*, ed. by T. Riste (Plenum, New York 1977)
4.76 G. Gehring: accompanying volume
4.77 M. Born, K. Huang: *Dynamical Theory of Crystal Lattices* (Oxford University Press, Oxford 1954)
4.78 M.E. Mullen, B. Lüthi, P.S. Wang, E. Bucher, L.D. Longinotti, J.P. Maita, H.R. Ott: Phys. Rev. B*10*, 186 (1974)

4.79 G. Gorodetsky, A. Shaulov, V. Volterra, J. Makovsky: Phys. Rev. *B13*, 1205 (1976)
4.80 B. Lüthi: AIP Conf. Proc. *34*, 7 (1976)
4.81 P. Fulde: In *Handbook on the Physics and Chemistry of Rare Earths*, ed. by K.A. Gschneidner, L. Eyring (North-Holland, Amsterdam 1978) Chap.17
4.82 M.J. Sablik, H.H. Teitelbaum, P.M. Levy: AIP Conf. Proc. *10*, 548 (1972)
4.83 S.J. Allen: Phys. Rev. *167*, 492 (1968)
4.84 M. Kataoka, J. Kanamori: J. Phys. Soc. Jpn. *32*, 113 (1972)
4.85 P.M. Levy: J. Phys. *C6*, 3545 (1973)
4.86 Y. Kino, S. Miyahara: J. Phys. Soc. Jpn. *21*, 2732 (1966)
4.87 Y. Kino, B. Lüthi, M.E. Mullen: J. Phys. Soc. Jpn. *33*, 687 (1972)
Y. Kino, B. Lüthi, M.E. Mullen: Solid State Commun. *12*, 275 (1973)
4.88 H. Terauchi, M. Mori, Y. Yamada: J. Phys. Soc. Jpn. *32*, 1049 (1972)
4.89 S. Hirotsu: J. Phys. *C10*, 967 (1977)
4.90 C.J. Kroese, W.J.A. Maaskant: Chem. Phys. *5*, 224 (1974)
4.91 Y. Ishikawa, Y. Syono: Phys. Rev. Lett. *26*, 1335 (1971)
4.92 J.R. Sandercock, S.B. Palmer, R.J. Elliott, W. Hayes, S.R.P. Smith, A.P. Young: J. Phys. *C5*, 3126 (1972)
4.93 R.L. Melcher, E. Pytte, B.A. Scott: Phys. Rev. Lett. *31*, 307 (1973)
4.94 G. Gorodetsky, B. Lüthi, B.M. Wanklyn: Solid State Commun. *9*, 2157 (1971)
4.95 R.T. Harley, R.M. Macfarlane: J. Phys. *C8*, L 451 (1975)
4.96 F. Levy: Phys. Kondens. Mater. *10*, 85 (1969)
4.97 B. Lüthi, M.E. Mullen, K. Andres, E. Bucher, J.P. Maita: Phys. Rev. *B8*, 2639 (1973)
4.98 P. Morin, J. Rouchy, D. Schmitt: Phys. Rev. *B17*, 3684 (1978)
B. Lüthi, R. Sommer, P. Morin: J. Magn. Magn. Mater. *13*, 198 (1979)
K.M. Leung, D.L. Huber, B. Lüthi: J. Appl. Phys. *50*, 1831 (1979)
4.99 R.J. Birgeneau, J.K. Kjems, G. Shirane, L.G.v. Uitert: Phys. Rev. *B10*, 2512 (1974)
4.100 K. Andres, P.S. Wang, Y.H. Wong, B. Lüthi, H.R. Ott: AIP Conf. Proc. *34*, 222 (1976)
4.101 O.G. Brandt, C.T. Walker: Phys. Rev. Lett. *18*, 11 (1967)
4.102 J. Faber, Jr., G.H. Lander, B.R. Cooper: Phys. Rev. Lett. *35*, 1770 (1975)
4.103 B.W. Batterman, C.S. Barrett: Phys. Rev. *145*, 296 (1966); Phys. Rev. Lett. *13*, 390 (1964)
4.104 L.R. Testardi: In *Physical Acoustics*, Vol.X, ed. by W.P. Mason, R.N. Thurston (Academic, New York 1973) p.193
4.105 M. Weger, I.B. Goldberg: In *Solid State Physics*, Vol.28, ed. by H. Ehrenreich, F. Seitz, D. Turnbull (Academic, New York 1973) p.1
4.106 W.E. Blumberg, J. Eisinger, V. Jaccarino, B.T. Matthias: Phys. Rev. Lett. *5*, 149 (1960)
4.107 A.M. Clogston, V. Jaccarino: Phys. Rev. *121*, 1357 (1961)
4.108 J. Labbe, J. Friedel: J. Phys. (Paris) *27*, 153, 303, 708 (1966)
4.109 R.W. Cohen, G.D. Cody, J.J. Halloran: Phys. Rev. Lett. *19*, 840 (1967)
4.110 L.R. Testardi, T.B. Bateman: Phys. Rev. *154*, 402 (1967)
4.111 K.R. Keller, J.J. Hanak: Phys. Rev. *154*, 628 (1967)
4.112 W. Rehwald, M. Rayl, R.W. Cohen, G.D. Cody: Phys. Rev. *B6*, 363 (1972)
W. Rehwald: Phys. Lett. *27A*, 287 (1968)
4.113 J.D. Axe, G. Shirane: Phys. Rev. *B8*, 1965 (1973)
4.114 L.F. Mattheiss: Phys. Rev. *B12*, 2161 (1975); *138*, A 112 (1965)
4.115 L.P. Gorkov, O.N. Dorokhov: J. Low Temp. Phys. *22*, 1 (1976)
4.116 R.N. Bhatt, W.L. McMillan: Phys. Rev. *B14*, 1007 (1976)
4.117 R.N. Bhatt: Phys. Rev. *B16*, 1915 (1977)
4.118 T.-K. Lee, J.L. Birman, S.J. Williamson: Phys. Rev. Lett. *39*, 839 (1977)
4.119 W. Dieterich, G. Stollhof: Z. Phys. *266*, 185 (1974)
4.120 J.D.N. Cheeke, H. Mallie, S. Roth, B. Seeber: Solid State Commun. *13*, 1567 (1973)
4.121 S. Roth, W. Rehwald, H. Mallie: Solid State Commun. *22*, 177 (1977)
4.122 H. Ihrig, D.T. Vigren, J. Kübler, S. Methfessel: Phys. Rev. *B8*, 4525 (1973)
4.123 W. Assmus, R. Takke, R. Sommer, B. Lüthi: J. Phys. *C11*, L 575 (1978)
4.124 R.D. Parks (ed.): *Valence Instabilities and Related Narrow-Band Phenomena* (Plenum, Rochester 1976)

4.125 T. Penney, R.L. Melcher, F. Holtzberg, G. Güntherodt: AIP Conf. Proc. *29*, 392 (1976)
4.126 M.C. Croft, R.D. Parks: In *Valence Instabilities and Related Narrow-Band Phenomena*, ed. by R.D. Parks (Plenum, Rochester 1976) p.455
4.127 H.R. Ott, B. Lüthi, P.S. Wang: In *Valence Instabilities and Related Narrow-Band Phenomena*, ed. by R.D. Parks (Plenum, Rochester 1976) p.289
4.128 B. Batlogg, H.R. Ott, E. Kaldis, W. Thöni, P. Wachter: Phys. Rev. B*19*, 247 (1979)
4.129 D.J. Gunton, G.A. Saunders: Solid State Commun. *14*, 865 (1974)
M.R. Madhava, G.A. Saunders, R.C. Draper: Phys. Lett. *68*A, 67 (1978)
4.130 Y. Marukani: J. Phys. Soc. Jpn. *33*, 1350 (1972)
4.131 M. Mori, Y. Yamada, G. Shirane: Solid State Commun. *17*, 127 (1975)
4.132 M.W. Finnis, V. Heine: J. Phys. F*4*, 960 (1974)
M.W. Finnis: J. Phys. F*4*, 969 (1974)
4.133 W. Rehwald: Phys. Kondens. Mater. *14*, 21 (1971); Solid State Commun. *8*, 607, 1483 (1970)
4.134 R.O. Bell, C. Rupprecht: Phys. Rev. *129*, 90 (1963)
4.135 B. Lüthi, T.J. Moran: Phys. Rev. B*2*, 1211 (1970)
4.136 K. Fossheim, B. Berre: Phys. Rev. B*5*, 3292 (1972)
4.137 W. Rehwald: Solid State Commun. *21*, 667 (1977)
4.138 B. Okai, J. Yoshimoto: J. Phys. Soc. Jpn. *39*, 162 (1975)
4.139 D. Bäuerle, W. Rehwald: Solid State Commun. *27*, 1343 (1978)
4.140 J.M. Courdille, J. Dumas: Solid State Commun. *7*, 1623 (1969)
4.141 E.R. Domb, H.K. Schurmann, T. Mihalisin: Phys. Rev. Lett. *36*, 1191 (1976)
4.142 L.M. Reshchikova, V.I. Zinenko, K.S. Aleksandrov: Sov. Phys. Solid State *11*, 2893 (1970)
4.143 K. Gesi, J.D. Axe, G. Shirane, A. Linz: Phys. Rev. B*5*, 1933 (1972)
4.144 J.N. Bierly, L. Muldawer, O. Beckman: Acta Metall. *11*, 447 (1963)
4.145 E.F. Steigmeier, G. Harbeke: Solid State Commun. *8*, 1275 (1970)
4.146 G.S. Pawley, W. Cochran, R.A. Cowley, G. Dolling: Phys. Rev. Lett. *17*, 753 (1966)
4.147 M. Iizumi, Y. Hamaguchi, K.F. Komatsubara, Y. Kato: J. Phys. Soc. Jpn. *38*, 443 (1975)
4.148 W. Rehwald, G.K. Lang: J. Phys. C*8*, 3287 (1975)
4.149 T. Seddon, J.M. Farley, G.A. Saunders: Solid State Commun. *17*, 55 (1975)
4.150 I. Hatta, W. Rehwald: J. Phys. C*10*, 2075 (1977)
4.151 R. Clarke: Phys. Rev. B*18*, 4920 (1978)
4.152 J.V. Atanasoff, Ph.J. Hart: Phys. Rev. *59*, 85, 97 (1941)
4.153 E. Kammer, T.E. Perdue, H.F. Frissel: J. Appl. Phys. *19*, 265 (1947)
4.154 U.T. Höchli, J.F. Scott: Phys. Rev. Lett. *26*, 1627 (1971)
4.155 J.P. Bachheimer, G. Dolino: Phys. Rev. B*11*, 3195 (1975)
4.156 U.T. Höchli: Phys. Rev. B*6*, 1814 (1972)
4.157 J.M. Courdille, R. Deroche, J. Dumas: J. Phys. (Paris) *36*, 891 (1975)
4.158 B. Dorner, J.D. Axe, G. Shirane: Phys. Rev. B*6*, 1950 (1972)
4.159 I.J. Fritz: Phys. Lett. *51*A, 219 (1975)
4.160 D.E. Cox, S.M. Shapiro, R.A. Cowley, M. Eibschütz, H.J. Guggenheim: Phys. Rev. B*19*, 5754 (1979)
4.161 A.I. Larkin, D.E. Kmel'nitskii: JETP *29*, 1123 (1969)
4.162 T. Nattermann: Phys. Status Solidi B*85*, 291 (1978)
4.163 D. Stauffer: Ferroelectrics *18*, 199 (1978)
4.164 B.A. Strukov, K.A. Minaeva, N.M. Shirina, V.I. Teleshevskii, S.K. Khanna: Izv. Akad. Nauk. SSSR, Ser. Fiz. *39*, 180 (1975)
4.165 B.A. Strukov: Ferroelectrics *12*, 97 (1976)
4.166 I. Todo: J. Phys. Soc. Jpn. *39*, 1538 (1975)
4.167 K.A. Minaeva, A.P. Levanyuk, B.A. Strukov, V.A. Koptsik: Sov. Phys. Solid State *9*, 950 (1967)
4.168 K. Kawasaki: Phys. Lett. *29*A, 406 (1969)
4.169 K. Ota, Y. Ishibashi, Y. Takagi: J. Phys. Soc. Jpn. *29*, 1545 (1970)
4.170 I. Hatta, Y. Shimizu, K. Hamano: J. Phys. Soc. Jpn. *44*, 1887 (1978)
I. Hatta, M. Hanami, K. Hamano: J. Phys. Soc. Jpn. *48*, 160 (1980)
4.171 I. Todo, I. Tatsuzaki: J. Phys. Soc. Jpn. *31*, 1479 (1971)
4.172 I. Todo, I. Tatsuzaki: Phys. Status Solidi A*23*, 591 (1974)

4.173 V.I. Samulionis, V.F. Kunigelis: Fiz. Tver. Tela *13*, 740 (1971) [English transl.: Sov. Phys. - Solid State *13*, 611 (1971)]
4.174 I. Todo, T. Tatsuzaki: J. Phys. Soc. Jpn. *37*, 1477 (1974)
4.175 Sh. Kashida, I. Hatta, A. Ikushima, Y. Yamada: J. Phys. Soc. Jpn. *34*, 997 (1973)
4.176 R. Clarke, L. Benguigui: J. Phys. C*10*, 1963 (1977)
4.177 H. Kameyama, Y. Ishibashi, Y. Takagi: J. Phys. Soc. Jpn. *33*, 861 (1972); *36*, 614 (1974); *46*, 566 (1979)
4.178 G.R. Barsch, L.J. Bonczar, R.E. Newnham: Phys. Status Solidi A*29*, 241 (1975)
4.179 E.P. Maishchik, B.A. Strukov, E.V. Sinyakov, K.A. Minaeva, V.G. Monya: Fiz. Tver. Tela *19*, 335 (1977) [English transl.: Sov. Phys. - Solid State *19*, 193 (1977)]
4.180 P.A. Fleury, K.B. Lyons: Phys. Rev. Lett. *37*, 1088 (1976)
4.181 C.W. Garland: In *Physical Acoustics*, Vol.VII, ed. by W.P. Mason, R.N. Thurston (Academic, New York 1970) p.129
4.182 R.A. Cowley, J.D. Axe, M. Iizumi: Phys. Rev. Lett. *36*, 806 (1976)
4.183 A. Hüller: Z. Phys. *254*, 456 (1972)
4.184 Y. Yamada, M. Mori, Y. Noda: J. Phys. Soc. Jpn. *32*, 1560 (1972)
4.185 C.W. Garland, R. Renard: J. Chem. Phys. *44*, 1130 (1966)
C.W. Garland, B.B. Weiner: Phys. Rev. B*3*, 1634 (1971)
4.186 C.W. Garland, J.D. Baloga: Phys. Rev. B*16*, 331 (1977)
4.187 D.J. Bergmann, B.I. Halperin: Phys. Rev. B*13*, 2145 (1976)
4.188 C.W. Garland, C.K. Choo: Phys. Rev. B*8*, 5143 (1973)
4.189 T. Tiedje, R.R. Haering, M.H. Jericho, W.A. Roger, A. Simpson: Solid State Commun. *23*, 713 (1977)
4.190 H. Doi, H. Nagasawa, T. Ishiguro, S. Kagoshima: Solid State Commun. *24*, 729 (1977)
4.191 M. Barmatz, L.R. Testardi, A.F. Garito, A.J. Heeger: Solid State Commun. *15*, 1299 (1974)
4.192 M. Barmatz, L.R. Testardi, F.J. DiSalvo: Phys. Rev. B*12*, 4367 (1975)
4.193 M.S. Skolnick, S. Roth, H. Alms: J. Phys. C*10*, 2523 (1977)
4.194 R. Geick, K. Strobel: J. Phys. C*10*, 4221 (1977)
T. Goto, B. Lüthi, R. Geick, K. Strobel: J. Phys. C*12*, L 303 (1979)
4.195 S. Geller (ed.): *Solid Electrolytes*, Topics in Applied Physics, Vol.21 (Springer, Berlin, Heidelberg, New York 1977)
4.196 M.B. Salamon (ed.): *Physics of Superionic Conductors*, Topics in Current Physics, Vol.15 (Springer, Berlin, Heidelberg, New York 1979)
4.197 L.J. Graham, R. Chang: J. Appl. Phys. *46*, 2433 (1975)
4.198 M. Nagao, T. Kaneda: Phys. Rev. B*11*, 2711 (1975)
4.199 M. Maeda, T. Ikeda: J. Phys. Soc. Jpn. *42*, 1931 (1977)
4.200 Y. Makita, F. Sakurei, T. Osaka, I. Tatsusaki: J. Phys. Soc. Jpn. *42*, 518 (1977)
4.201 S. Haussühl, J. Eckstein, R. Recker, F. Wallrafen: Acta Crystallogr. A*33*, 847 (1977)
4.202 W.F. Love, H.D. Hochheimer, M.W. Anderson, R.N. Work, C.T. Walker: Solid State Commun. *23*, 365 (1977)
4.203 E. Bucher, J.P. Maita, G.W. Hull, Jr., L.D. Longinotti, B. Lüthi, P.S. Wang: Z. Phys. B*25*, 41 (1976)
4.204 U.T. Höchli: Solid State Commun. *8*, 1487 (1970)
4.205 J.M. Courdille, J. Dumas: J. Phys. (Paris) *36*, L5 (1975)
4.206 H. Kameyama, Y. Ishibashi, Y. Takagi: J. Phys. Soc. Jpn. *38*, 1706 (1975)
4.207 H. Kameyama, Y. Ishibashi, Y. Takagi: J. Phys. Soc. Jpn. *35*, 1450 (1973)
4.208 K.M. Leung, D.L. Huber: Phys. Rev. Lett. *42*, 452 (1979)
4.209 T. Yagi, M. Cho, Y. Hidaka: J. Phys. Soc. Jpn. *46*, 1957 (1979)
4.210 S. Kudo, T. Ikeda: Jpn. J. Appl. Phys. *19*, L45 (1980)
4.211 W. Rehwald, A. Vonlanthen, J.K. Krüger, R. Wallerius, H.G. Unruh: J. Phys. C *13*, 3823 (1980)
4.212 S. Hirotsu, K. Toyota, K. Hamano: J. Phys. Soc. Jpn. *46*, 1389 (1979)
4.213 H. Hoshizaki, A. Sawada, Y. Ishibashi: J. Phys. Soc. Jpn. *47*, 341 (1979)
4.214 S. Kudo, T. Hihita: J. Phys. Soc. Jpn. *45*, 1775 (1978)
4.215 S. Kashida: J. Phys. Soc. Jpn. *45*, 1874 (1978)

Additional References with Titles

H. Boppart, A. Treindl, P. Wachter, S. Roth: First observation of a negative elastic constant in intermediate valent TmSe, Solid State Commun. *35*, 483 (1980)

S. Brühl: Dynamics of cooperative Jahn-Teller T-systems: elastic properties, Z. Phys. B*37*, 231 (1980)

K. Fossheim, R.M. Holt: Critical dynamics of sound in $KMnF_3$, Phys. Rev. Lett. *45*, 730 (1980)

W. Henkel, J. Pelzl, K.-H. Höck, H. Thomas: Elastic constants and softening of acoustic modes in A_2MX_6-crystals observed by brillouin scattering, Z. Phys. B*37*, 321 (1980)

K. Knorr, B. Renker, W. Assmus, B. Lüthi, R. Takke, H.J. Lauter: $LaAg_xIn_{1-x}$: phonon dispersion, elastic constants and structural instability, Z. Phys. B*39*, 151 (1980)

K.M. Leung, D.L. Huber: Low-frequency dynamics in cooperative Jahn-Teller systems, Phys. Rev. B*19*, 5483 (1979)

J. Maetz, M. Müllner, H. Jex, W. Assmus, R. Takke: $LaAg_xIn_{1-x}$: crystal structures determined by neutron diffraction, Z. Phys. B*37*, 39 (1980)

Subject Index

Dynamic of Solids and Liquids by Neutron Scattering

Editors: S. W. Lovesey, T. Springer
1977. 156 figures, 15 tables. XI, 379 pages
(Topics in Current Physics, Volume 3) ISBN-13: 978-3-642-81533-1

Contents:
S. W. Lovesey: Introduction. – *H. G. Smith, N. Wakabayashi:* Phonons. – *B. Dorner, R. Comès:* Phonons and Structural Phase Transformations. – *J. W. Withe:* Dynamics of Molecular Crystals, Polymers, and Absorbed Species. – *T. Springer:* Molecular Rotations, and Diffusion in Solids, in Particular Hydrogen in Metals. – *R. D. Mountain:* Collective Modes in Classical Monoatomic Liquids. – *S. W. Lovesey, J. M. Loveluck:* Magnetic Scattering.

Neutron Diffraction

Editor: H. Dachs
1978. 138 figures, 32 tables. XIII, 357 pages
(Topics in Current Physics, Volume 6) ISBN-13: 978-3-642-81533-1

Contents:
H. Dachs: Principles of Neutron Diffraction. – *J. B. Hayter:* Polarized Neutrons. – *P. Coppens:* Combining X-Ray and Neutron Diffraction: The Study of Charge Density Distributions in Solids. – *W. Prandl:* The Determination of Magnetic Structures. – *W. Schmatz:* Disordered Structures. – *P.-A. Lindgård:* Phase Transitions and Critical Phenomena. – *G. Zaccaï:* Application of Neutron Diffraction to Biological Problems. – *P. Chieux:* Liquid Structure Investigation by Neutron Scattering. – *H. Rauch, D. Petrascheck:* Dynamical Neutron Diffraction and Its Application.

Physics of Superionic Conductors

Editor: M. B. Salamon
1979. 101 figures, 13 tables. XII, 255 pages
(Topics in Current Physics, Volume 15) ISBN-13: 978-3-642-81533-1

Contents:
M. B. Salamon: Introduction. – *J. B. Boyce, T. M. Hayes:* Structure and Its Influence on Superionic Conduction. EXAFS Studies. – *S. M. Shapiro, F. Reidinger:* Neutron Scattering Studies of Superionic Conductors. – *H. U. Beyeler, P. Brüesch, L. Pietronero, W. R. Schneider, S. Strässler, H. R. Zeller:* Statics and Dynamics Lattice Gas Models. – *M. J. Delaney, S. Ushioda:* Light Scattering in Superionic Conductors. – *P. M. Richards:* Magnetic Resonance in Superionic Conductors. – *M. B. Salamon:* Phase Transitions in Ionic Conductors. – *T. Geisel:* Continous Stochastic Models. – Additional References with Titles. – Subject Index.

H. Bilz, W. Kress

Phonon Dispersion Relations in Insulators

1979. 162 figures in 271 separate illustrations. VIII, 241 pages
(Springer Series in Solid-State Sciences, Volume 10)
ISBN-13: 978-3-642-81533-1

Contents:
Summary of Theory of Phonons: Introduction. Phonon Dispersion Relations and Phonon Models. – Phonon Atlas of Dispersion Curves and Densities of States: Rare-Gas Crystals. Alkali Halides (Rock Salt Structure). Metal Oxides (Rock Salt Structure). Transition Metal Compounds (Rock Salt Structure). Other Cubic Crystals (Rock Salt Structure). Cesium Chloride Structure Crystals. Diamond Structure Crystals. Zinc-Blende Structure Crystals. Wurtzite Structure Crystals. Fluorite Structure Crystals. Rutile Structure Crystals. ABO_3 and ABX_3 Crystals. Layered Structure Crystals. Other Low-Symmetry Crystals. Molecular Crystals. Mixed Crystals. Organic Crystals. – References. – Subject Index.

Springer-Verlag
Berlin
Heidelberg
New York